GLOSSARY OF
MARINE
TECHNOLOGY
TERMS

The reasons a ship is called 'She'

There is always a great deal of bustle about her, there is usually a gang of men around her, she has waists and stays, she takes a lot of paint to keep her looking good, she shows her topsides, hides her bottom, and when coming into port always heads for the buoys.

Sir John Martin
Lieutenant Governor of Guernsey

See *Annual Report R.I.N.A.* 1976, Vol. 118, p. LIX

GLOSSARY OF MARINE TECHNOLOGY TERMS

Published in association
with the Institute of
Marine Engineers

HEINEMANN : LONDON

William Heinemann Ltd
10 Upper Grosvenor Street, London W1X 9PA

LONDON   MELBOURNE   TORONTO
JOHANNESBURG   AUCKLAND

First published 1980
© The Institute of Marine Engineers 1980
434 908401

Printed by
Willmer Brothers Limited,
Birkenhead, Merseyside

List of Figures

		page
Figure 1	Rudder Types	10
Figure 2	Terms used in naval architecture (i)	12
Figure 3	Union purchase rig	14
Figure 4	Body plan of cargo ship	16
Figure 5	Terms used in naval architecture (ii)	21
Figure 6	Bulk carrier, mid-ship section	21
Figure 7	Cavitation tunnel tests on 4-bladed propeller	26
Figure 8	Transverse metacentre	28
Figure 9	Duct keel	48
Figure 10	Floodable length	58
Figure 11	Keel	86
Figure 12	Load lines	94
Figure 13	Primary mode	106
Figure 14	Modes of vertical vibration	106
Figure 15	Parallel middle body	132
Figure 16	Steam temperatures	153
Figure 17	Wheatstone bridge	176

Introduction

This book includes many of the words which appear in the Department of Trade's Examinations for Certificates of Competancy. Terminology used by deck, marine engineering and electrical engineering officers is included.

English is the common language of the International Shipping Industry and it is hoped that the book will be of use to shipping personnel in third world countries who are developing their own merchant fleets as well as those studying for examinations.

A

Able seaman

Experienced seaman competent to perform usual and customary duties on deck.

A-bracket

Certain fast twin-screw ships and most warships have propeller shafts extending outside the hull of the ship forward of the stern post. Such shafts are supported by a bearing in an A-bracket near the propeller, so called because the bracket resembles the letter A lying on its side.

Abrasion

When used with reference to gearing is the action of scraping or scoring of the gear teeth, caused by solid particles in the lubricating oil and the sliding effect of one tooth over another.

Abrasion in diesel engines

Common wear mechanism characterized by fairly regularly spaced grooves running in rubbing direction. Usually more prominent near top dead centre.

Abrasive wear

See *Gear tooth damage.*

Abrasives

Used mainly in grinding, honing and polishing. Classified as: (1) natural sandstone, quartz, emery, corundum, natural diamonds. (2) synthetic silicon carbide, aluminium oxide, diamonds.

Accommodation

Space in ship for sleeping, mess rooms, wash rooms and recreation.

Accommodation ladder

Portable ladder attached to platform at ship's side and which can be positioned to give access to ship from water or shore.

Accumulator

(1) Electric storage battery for which there are British Standards applicable for lead-acid and alkaline types. (2) Kierselbach accumulator of thermal storage enables a boiler to cope with heavy demands for steam. (3) Accumulators can be used to store liquid and gases at constant pressure, acting as a reservoir.

Addendum

That part of the working surface of a gear tooth

1

which is towards the tip of the tooth above the pitch line.

Additives

Chemicals added as minor components to fuel oil or lubricating oil. Used in the fuel oil to improve combustion and prevent the formation of deposits such as corrosive oxides of vanadium which may collect on superheater tubes. In the case of lubricating oils, chemicals may be added to affect either the chemical or the physical characteristics. Included to improve the chemical characteristics; anti-oxidants, wear-reducing agents, detergent/dispersant additives, high alkalinity additives, and anti-bacterial additives may be added. To improve the physical characteristics chemicals may be added to prevent foaming, reduce the pour point, improve the viscosity index, and prevent emulsification.

Adhesives

Used to unite materials so that they adhere more or less permanently. An adhesive should, in general, 'wet' the surfaces being bonded and on solidification should not disrupt the bond. There are many kinds of adhesives, both organic and inorganic.

Adiabatic

Change of condition in which no heat is added or subtracted from a gas and during which no losses – due to friction and eddies – occur. Law of expansion is:
PV^n = a constant
 n = 1.4 for air at normal temperatures.

Admiralty constant

This was intended as a means of estimating the power requirements and comparing the performance of ships. After trial of an earlier form based on immersed midship section area, the constant was defined as:

$$\frac{\text{speed}^3 \times \text{displacement}^{2/3}}{I.H.P}$$

It is assumed that resistance varies as speed^2 and the propulsive efficiency (i.e. *E.H.P./I.H.P.*) remains constant. Unfortunately, though of use when considering relatively slow speed ships, large variations in its value occur with length and even with speed for the same vessel. For example, the value may range from 20 for a motor dinghy to 800 for a fast skimming craft; and from 250 at cruising speed to 150 at full speed for the same ship. For derivation, relationship to other coefficients and information on usage, *see* MUCKLE, W. *Naval Architecture for Marine Engineers* (Newnes and Butterworth 1975), BARNABY, K. C. *Basic Naval Architecture* (Hutchinson 1967) sections 118 and 192 and BAKER, G. S. *Ship Design, Resistance and Screw Propulsion* (Journal of Commerce, Liverpool, 1951) vol. 1, ch. 1. (Both the latter are currently out of print but can be obtained in libraries.)

Advance

Distance travelled by centre of gravity of a ship in a direction parallel to original course after the instant the rudder is put over.

After burning

(1) Combustion continued in internal combustion engines after exhaust ports or valves are opened resulting in flame in exhaust system which can sometimes ignite carbon or oil deposits. Normally due to faulty injectors or lack of compression but deliberately promoted in petrol engines to reduce harmful exhaust emissions. (2) Combustion in uptake from a boiler resulting from burner faults. (3) Fuel injected into jet engine exhaust to give additional thrust when aircraft is taking off.

After peak

Compartment at stern, abaft aftermost watertight bulkhead.

Air cooler

Consists of battery of finned tubes, material depending on the cooling medium. When warm humid air is exposed to the chilled surfaces of the cooler and comes into contact with the tubes or fins it is cooled to a mean surface temperature. If this is below its dew-point, moisture is deposited.

Air ejectors

Used to extract air from surface condensing plants. Air and vapour from the condenser is sucked by a steam jet ejector out of the condenser and cooled, the water condensate being returned to the system and the air to atmosphere.

3

Air heaters

Air may be heated through the medium of steam, water, electricity or gas. Steam or water heaters consist of tube banks – containing the heating medium – around which air passes. The steam is at a relatively low pressure. Electric air heaters are convenient for small systems. Gas air heaters have considerable industrial application.

Air-lock

(1) Compartment with air-tight door at each end. (2) Bubble of air in a pipe stopping flow of liquid.

Alkalinity

Extent to which a solution is alkaline – *see* pH value. The treatment of raw water to make it fit for boiler feeding depends upon the chemical analysis of the water concerned. The scale forming properties of water are due to its hardness. The total hardness is made up of 'temporary hardness' and 'permanent hardness'. Temporary hardness due to the presence of bicarbonates of calcium and magnesium in solution can be removed by boiling. Temporary hardness is now called 'alkalinity'.

Alkyd Resins

Tough, hard, durable products having excellent adhesion to most surfaces. Paints based on alkyd media are quick drying, of better durability and improved appearance. Used extensively in primers and undercoats.

Allen key

Cranked hexagonal bar for turning socket head screws.

Alternating current (a.c.)

Current in response to an alternating voltage, that reverses at regular intervals of time and has alternatively positive and negative values. Each complete reversal is termed a cycle. The number of cycles per second is termed the frequency.

Alternator

Machine that generates alternating voltage when its rotating portion is driven by motor.

Aluminium killed steel

During the steel-making process, deoxidation is necessary to prevent bubbling during pouring ('being lively'). Aluminium added to molten steel takes up oxygen and 'kills' the steel.

Ambient temperature Surrounding atmospheric temperature.

Amidships Mid-way between forward and after perpendiculars. Denoted by $\mathfrak{L}$. Term used when rudder is in the fore and aft or central position. See *Figure 5*.

Ammeter Instrument for measuring electric current in amperes.

Ammonia Noxious gas with a pungent smell which is very soluble in water giving an alkaline solution. Its chemical formula is NH_3.

Amplifier Device capable of increasing magnitude or power level of a physical quantity. Most amplifiers are electronic but pneumatic or hydraulic units may be used.

Analogue Refers to type of instrument which converts numbers, voltages or readings into physical quantities, e.g. slide rule converts a number into a length; car speedometer converts revolutions into speed and an analogue computer converts input into an electric current. For a given accuracy an analogue instrument tends to be more bulky than digital equivalent, e.g. digital car speedometer is smaller and more accurate than existing analogue pointer type.

Anchor Implement by which a ship is rendered stationary. The main anchors carried aboard ship are of two types, namely stock anchors and stockless anchors. Anchors and cables are inspected at annual docking survey.

Anode Positive electrode of electrolytic cell. See *Cathode*.

Anode (sacrificial) In a system of cathodic protection, sacrificial anodes corrode naturally and in so doing protect from corrosion the material to which they are attached which is usually steel.

5

Sacrificial anodes mounted on the hull of a ship are made of zinc or aluminium alloy, the composition of which is strictly controlled.

Anti-fouling paint

Paint which is applied to the wetted surface of a ship to prevent the adherence and growth of biological organisms, such as slime, weed and shell. Such growth leads to increased resistance, requiring additional power and fuel to maintain same speed. The composition includes poisons which leach out slowly into the water.

Aperture

Space between rudder post and propeller post for propeller.

Arc welding

Process for joining metals by striking an electric arc between the parts to be joined, thereby causing melting and fusion. Many welding processes use an electric arc; among the most common are manual metal arc (MMA), submerged arc and metal inert gas (MIG).

Argon

An element; an inert gas present in small quantity in the earth's atmosphere. It is used in electric light bulbs and for shielding the molten metal from oxidation in argon arc welding. Chemical symbol – Ar.

Armature (relay)

The moving element that contributes to the designed response of the relay and which usually has associated with it a part of the relay contact assembly.

Armature (rotating machine)

The member of an electrical machine in which an alternating voltage is generated by virtue of relative motion with respect to a magnetic field.

Articulation

Any system of components such as gears or levers in which the sharing or equalization of loads is effected by proportioning the leverages in relation to the fulcrum. An epicyclic train of gears in whch the input torque is applied via the planet cage, the planet gears acting as articulated levers to share the tooth loads between sun wheel and annulus ring. This principle is also the basis of a differential gear train. The principle of articulation is also used to equalize the loads on the wheels of multi-wheeled vehicles and lorries.

Asbestos

Fibrous mineral of calcium and magnesium silicates that can be woven into incombustible fabrics and used for thermal insulation of pipes, etc. Certain types are a dangerous health hazard and operators fitting or removing lagging must wear breathing apparatus and protective clothing.

ASDIC

Echo sounding device for detecting underwater objects. Initials of 'Anti-Submarine Detection Investigation Committee'. Modern word is SONAR.

Asperities

Excrescences on a surface producing roughness. Usually applied to the surface of steel after blast cleaning $(q.v.)$ where the use of coarse grit may produce peaks which penetrate the subsequently applied paint coating.

Astern power

Power available for driving a ship astern. In direct drive diesel machinery power available is 100% of that available for ahead movement; in turbine driven ships, as a rule, only a proportion of ahead power is provided by separate stages of turbine blades. Propeller efficiency is considerably lower when ship is going astern.

Atmospheric Condenser

Type of condenser with cooling tower using the atmosphere as the cooling medium. Tower is designed to provide natural air draught and water falls over cooling pipes situated in air flow, heat being removed as water evaporates. Mainly used in power stations which have limited supplies of cooling water and where the large size of the cooling tower is acceptable. Another arrangement is to lead a steam drain coil into a tank of water, using the water to condense the steam and minimizing loss of steam to the atmosphere.

Atmospheric drain tank

The tank in the feed system of a steam plant to which condensed water from auxiliaries' heating coils etc. are led. The tank is at atmospheric pressure.

Atmospheric valve

(1) Valve in closed exhaust steam line which opens to atmosphere when pressure exceeds a certain limit. (2) Pressure/vacuum valve (Blundell Atmos Valve) fitted to tank top to maintain atmospheric pressure in tank. Valve vents excessive pressure and allows air to enter tank under vacuum conditions.

Atomization

Subdivision of a material into its smallest parts, particularly applied to liquids reduced to a fine spray, e.g. diesel engine injectors.

Attemporator

A heat exchanger used in the control of the final superheat temperature of the steam to the main engine.

Augmentor

A device which will increase or improve the operation or performance of some piece of equipment already in a system. Steam ejector fitted to steam reciprocating engine to assist reciprocating air pump when main engine is stopped.

Auto kleen strainers and filters

Proprietory strainer consisting of spiral wound wire on drum with small clearance between adjacent wires. Small kegs are fitted between each pair of wires which can be rotated manually or automatically to clean the strainer. Strainers fitted in pairs can be designed to change over automatically when a strainer becomes choked as the pressure drop across the strainer exceeds a certain limit, or automatically to clean the strainer, the displaced residue falling into a sump.

Auto-start valve

Fitted in the main air line to the starting air valves of a diesel engine. Operation of the starting lever causes the valve to open and air is then supplied to the air distributor and individual cylinder starting air valves. The valve will close once the starting lever is released.

Automatic control systems

Control system in which the value of a controlled or a related condition is compared

8

with a desired value, and a corrective action, dependent upon the deviation, is taken without the intervention of a human element. An automatic-control system includes a measuring unit, a controlling unit, a correcting unit and the plant being controlled. Mechanism which measures the value of a process variable and operates to limit the deviation of such variable from a desired value. The theory of automatic control systems is rather complex and has been the subject of extensive mathematical analysis. BS 1523 gives a Glossary of terms used in Automatic Controlling and Regulating Systems.

Automatic helmsman

See *Automatic pilot*

Automatic pilot

Many ships travel for long periods of time on a fixed course, the only deviations in course being those created by variations in tide, waves or wind. In such circumstances the automatic control system – automatic pilot – is most valuable. The system can sense the difference between the ordered course and the actual course and can cause the rudder to move to an angle proportional to this error.

Automatic voltage regulator (A.V.R.)

Device which operates to maintain the output voltage of a generator within predetermined limits, usually by controlling the excitation of the generator in response to load changes.

Automatic watchkeeper

Equipment incorporating monitoring and alarm surveillance. Facilities for alarm recording, trend recording, automatic log keeping, performance monitoring and automatic plant protection may also be provided. Print-out, e.g. on an automatic typewriter, can have any pre-programme format.

Automation

Development of highly automatic machinery or control systems which reduce or dispense with watchkeepers.

Autotransformer

Transformer consisting of one electrically continuous winding with one or more fixed movable taps so that part of the winding is common to both primary and secondary circuits.

Average adjuster	Skilled person who apportions the loss and expenditure between the interested parties involved in a maritime adventure or claim under a general average act. An expert on all aspects of marine insurance law and loss adjustments.
Axial	In line with, or pertaining to, an axis.

B

Back lash	Lost angular motion in a mechanical transmission system due to the working clearance between components. Excessive backlash in geared systems is generally caused by wear of the gear teeth and/or bearings.
Back pressure	Pressure on the exhaust side of an engine or system. The output or efficiency of the system will be reduced if the back pressure exceeds the designed value.
Balanced rudder	Rudder type in which a proportion of the

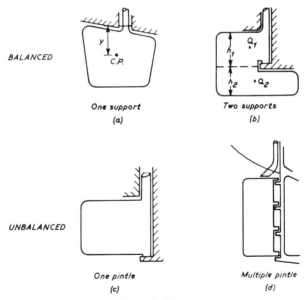

BALANCED

One support
(a)

Two supports
(b)

UNBALANCED

One pintle
(c)

Multiple pintle
(d)

Figure 1. Rudder types

rudder area – 25 to 30% – is forward of the axis of turning. The object is to reduce the operating torque required at the rudder stock since the effective centre of pressure acting on a balanced rudder is closer to the rudder stock axis. (Figure 1.)

Ball and roller bearings

Ball and roller bearings have low friction losses as the relative movement between two surfaces is accommodated by the rolling motion of intermediary case hardened spheres or cylinders. Such bearings are extensively used as journal or thrust bearings in shafting systems where cleanliness of the lubrication can be guaranteed. At higher load factors roller bearings are employed.

Ballast

Any solid mass or liquid placed in a ship (1) to increase the draught (2) to change the trim or to regulate the stability.

Ballast line

Piping system used to fill and empty ballast tanks.

Bareboat time charter

With this the entire responsibility for maintaining the ship in good repair and the manning devolves upon the charterer. Also the charterer during the charter must carry out all repairs necessary to maintain the vessel's class. In normal times this type of charter does not occur frequently as shipowners will not readily agree to the temporary transfer of management and control to charterers.

Barrel

Revolving cylinder on a machine. See also *Barrel (oil)*.

Barrel (oil)

Unit of volume used by the oil industry. 1 barrel = 42 U.S. gallons = 34.97 Imperial gallons. The term 'cask' is only applicable to alcoholic liquids.

Base line

For and aft reference line at the upper surface of the flat plate keel on the centre-line.

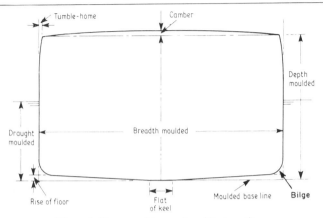

Figure 2. Terms used in naval architecture (i)

Beams	Athwartship steel rolled sections supporting the deck.
Bearings	Bearings are an essential part of all machines. Broadly, a bearing is an item common to two members of a mechanism such that they may move relatively to each other and transmit force from one to the other. Plain bearings are in three categories – journal, thrust and special. A journal bearing supports a rotating shaft and a thrust bearing is designed to withstand force along the axis of a shaft. Special bearings include ball or universal joints.
Beaufort scale	Wind force expressed numerically on a scale generally from 0 to 12. 0 Calm less then 1 kn 5 fresh breeze 17–21 kn 8 fresh gale 34–40 kn 12 hurricane above 65 kn
Bedplate	Structure forming base of machine.
Belfast bow	Name given to raked stem introduced by Harland and Wolff of Belfast. Gives large forecastle deck.
Bend test	Test for the ductility of a material, to reveal defects by bending a bar over a former of

specified radius related to the thickness of the bar.

b.h.p. Brake horsepower. The power output of a prime mover as measured by a dynamometer expressed in horsepower.
1 hp = 33,000 ft lb/min = 746 W.

Bilge (1) Figure 2. Curved portion, often circular, between bottom and side shell plating. (2) Drainage space within the ship.

Bilge cleaning The washing down of the bilge area of a ship to remove any pungent, unhygenic or dirty substances present. Bilges act as the collecting points for any free liquids in the holds and machinery space and if not regularly cleaned will rapidly corrode. Can be done manually or by special Tank Cleaning Vessels.

Bilge ejector A means of removing water from tank tops holds, etc. Water under pressure is passed through a nozzle, or venturi, the resulting jet entrains the water from the bilges and the mixture is discharged overboard.

Bilge keel External fin at round of bilge to reduce rolling. May extend outwards perpendicular to the ship's shell for, in some cases, up to 1 m. Also extends in the fore and aft direction for about one-half of the length of the ship.

Bilge strake Continuous horizontal fore and aft strip of plating from stem to stern in way of the bilge.

Bill of lading Ship Master's detailed receipt to consignor.

Binnacle Stand of wood or metal in which a compass is suspended. Top of binnacle protects compass from weather and reduces glare from lighting.

Bipod Mast Mast structure strong enough to dispense with supporting rigging. Suitable for use with heavy derricks and special derrick systems. (Figure 3.)

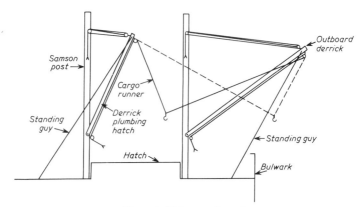

Figure 3. Union purchase rig

Bitt	Strong part of ship's structure, usually based on the keel and attached firmly to a main deck, to which a hawser, or warp, may be hitched when exceptionally heavy loads are applied, as when the vessel is towed. In small vessels this member is often termed the 'Samson Post' and the mooring chain or anchor cable is normally made fast to it, particularly when mooring in a tideway.
Bitumen	Mineral pitch or asphalt. Highly flammable viscous hydrocarbon which may be mixed with solvents to reduce softening temperature. Used for roadbuilding, weatherproofing and for protecting steel from corrosion. When carried as cargo, ships must be specially constructed with heated tanks isolated from the hull by double bottoms.
Blade, propeller	Screw propellers have two or more blades projecting from a boss. The surface of each blade viewed from aft is called the face and the other surface is the back.
Blading, impulse	Entire stage pressure drop occurs in stationary nozzles thus converting the internal energy of steam or combustion gas to kinetic energy. This steam jet does work on the rotor by impinging on a rotating row of blades reducing the velocity of the steam.

Blading, reaction
Type of blading fitted in turbines which progressively expands the steam as it passes through each row of fixed and moving blades. The extra kinetic energy released is then used up in driving the moving blades. Reaction blading gives a series of small pressure drops over the whole length of the turbine whereas with impulse blading a large pressure drop occurs in the turbine nozzles.

Blading, turbine
Radial projections on turbine rotors to convert pressure and/or kinetic energy of steam, water or combustion gas into mechanical work.

Blast cleaning
Process for cleaning steel by projecting grit, sand or shot at high velocity on to the surface. The best method to use for removal of millscale and rust prior to application of a coating to preserve the metal.

Block coefficient (C_b)
Ratio of the volume of displacement (V) to a given waterline and the volume of the circumscribing block of constant rectangular section having the same length (L), breadth (B) and draught (H) as the ship.

$$C_b = \frac{V}{LBH}$$

Blocks
In general, used to alter direction of a rope or chain or to gain a mechanical advantage. Two blocks are fitted to each derrick: one at the head known as the head block, gin block or cargo block; the other at the bottom known as a heel block. Types vary largely to suit different purposes. See also *Sheave blocks.*

Blow-down
Method of reducing dissolved solids in boiler water by opening blow-down cocks at bottom of boiler whilst maintaining correct water level in boiler guage glasses by introducing additional feed water to boiler. Same method used for reducing brine density in distillers.

Blower
Fitted to internal combustion engines to increase weight of air supplied to cylinders per stroke thereby increasing power output/ cylinder. Excess air may be provided to assist cylinder exhaust process. May be driven either

15

mechanically, electrically or by exhaust gas turbine.

Blower, LP　Low pressure air compressor for pneumatic tools, boiler cleaning, etc.

Blower, soot　Unit using steam jet for externally cleaning boiler tubes when steaming. Retractable when located in or close to the boiler furnace to avoid heat damage when not in use.

B.M.E.C.　British Marine Equipment Council. Trade association of British manufacturers of marine equipment.

b.m.e.p.　Brake mean effective pressure. The average pressure exerted on a piston during the working stroke.

Boatswain's chair　Wooden seat on which a man may be hoisted aloft or lowered over the ship's side in safety to carry out repairs or painting.

Body plan　Shows shape of transverse sections of the ship (Figure 4). See also *Lines plan*.

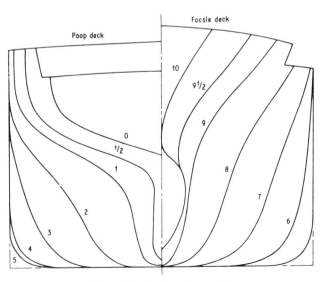

Figure 4. Body plan of cargo ship

Boiler combustion

The process of burning a fuel in a furnace to produce steam. Good combustion is important to avoid atmospheric pollution and fouling of the boiler tubes.

Boiler efficiency

Ratio of heat energy in steam delivered by boiler to heat energy supplied to boiler. Due to difficulties in measuring with sufficient accuracy (especially at sea) steam and oil fuel quantities, efficiency is usually computed by the method of losses. This subtracts heat losses from heat supplied and divides by the latter. Heat supplied is obtained from fuel characteristics and heat losses from flue gas conditions plus assumed radiation and other unaccountable losses. The latter is usually between 1 and 1.5% of the fuel Gross Combustion Value. Modern boilers with air preheaters can achieve efficiencies in excess to 90.5%.

Boiler water treatment

Provided to reduce corrosion and scale formation in boilers. Boiler water must be maintained in slightly alkaline condition by carefully controlled additions of chemicals.

Boil-off

Liquid evaporation due to heat transfer into the cargo or containment. Mainly applicable to LNG and LPG which are transported near their boiling points. Would not occur if perfect insulation was available.

Bollard

Large and firmly secured post of circular section for securing hawsers and mooring ropes. Often fitted in pairs on same base plate.

Bollard pull

Pull measured at bollard by, say, a tug at zero speed or at towing speed.

Bonjean curves

Curves of transverse sectional area of the ship drawn against a vertical scale or draught. Use of Bonjean curves enables immersed volume to be obtained for waterlines which are not parallel to base.

Boom

(1) Woodspar or steel tube used for discharging or loading cargo. (2) Spar for extending foot of sail. (3) Floating or moored

obstruction to restrict oil spillage or defend harbour entrance in wartime.

Boot-topping

Area of ship's side immediately above and below deep load line. Particularly susceptible to marine weed growth and often coated with anti-fouling paint specially formulated to prevent this type of growth. Term not applicable to cargo vessels which have large difference or distance between deep and light load lines

Bossings

Curved portion of ship's shell plating that surrounds and supports propeller shaft, offering less resistance than an 'A' bracket.

Bottle screw

Adjustable screw for tightening up stays and wires. Left-hand and right-hand threaded screws led into outer ends of a shroud and fitted with lock-nut or preventer.

Bow thruster

Reversible propeller mounted in a tunnel running through ship's hull to give movement athwartships in either direction and thus ease manoeuvring. Up to 15 thrusters may be fitted to a drill ship to enable accurate positioning above point on ocean bed.

Boyle's law

Volume of a given mass of gas at a constant temperature is inversely proportional to the pressure. Thus if V is the volume and P the pressure then at constant temperature $V \propto \frac{1}{P}$ or $PV = $ a constant.

Brazing

Process of joining metals by use of a copper based filler metal, the melting range of which is below that of either of the parent metals. The gap between the metals to be joined is such that molten filler metal is drawn in by capillary attraction. Capillary brazing used in manufacture of important pressure pipe systems as in warships and involves locating parts to be brazed in special jigs to ensure constant thickness of filler metal and maximum strength of joint.

Breadth, moulded

Measured at amidships and is maximum breadth over frames. See *Figure 2*.

Break bulk cargo	Miscellaneous goods packed in boxes, bales, cases, barrels or drums.
Breast hook	Triangular plate bracket joining port and starboard side stringers at stem, holding both sides of ship together.
Breast plate	Horizontal plate that connects shell plating at stem.
Bridge	Superstructure defined as erection above freeboard deck which extends to ship's side and gives clear view from which ship can be manoeuvred and navigated. Bridge structure in general located amidships or towards stern and forms important discontinuity in ship girder. In specialist ships may be situated nearer the bows. See also *Vessel – Bridge*.
Bridge gauge	Gauge for measuring wear-down on journal bearing with the shaft in place. Arch or bridge fitted over journal after top half bearing has been removed and clearance between bridge guage and journal measured.
Bridgewings	Open portion of the bridge extending from wheelhouse to side of vessel.
Bridle	Any fairly short length of rope secured at both ends.
Bridle gear	Metal cross-piece on end of accommodation ladder tackle to each end of which is attached a chain. Chains are shackled to ladder. Main function of bridle gear is to take weight of ladder.
Brine	(1) As a refrigerant is made by dissolving calcium chloride in fresh water and has a freezing point well below the desired temperatures of refrigerated compartments. (2) As seawater is evaporated in a distiller brine is formed which must be maintained at a specified density. Too high a density causes the coils to scale up quickly and too low a density wastes heat by discharging overboard too much brine.

19

Brine trap	Cylindrical chamber or pot fitted with inlet and outlet cocks or valves to enable samples of brine to be removed from distillers and refrigerator systems for testing the density.
Brinell hardness test	An indentation hardness test determined by applying a standard load to a standard size steel ball placed on the prepared surface of an article and measuring the diameter of the impression so produced. The hardness number is calculated by dividing the load applied (kg) by the surface area (mm^2) of the indentation or, in practice, by reference to tables.
British Marine Equipment Council (B.M.E.C.)	A trade association of manufacturers of British Marine Equipment.
Brittle fracture	Fracture of metal with very little plastic deformation. Ductile fracture implies slow fracture with significant elongation before fracture occurs. Propogation of brittle fracture is rapid, characterized by chevron shaped markings at fracture zone.
Brush gear	For support and adjustment of the brush used on electric motors, dynamos and engine starters, etc.
Bucket valve	Suction and discharge valves on the pump end of a reciprocating pump, normally consisting of sets of spring loaded plate valves.
Buckling	To crumple under longitudinal pressure.
Budgetary control	Method of controlling financial expenditure and income by placing limits and targets on the expenditure of all sections of an organization.
Bulbous bow	Protruding bow below waterline intended to reduce vessel's resistance to motion under certain circumstances (Figure 5). Was considered at one time to be favourable only for moderate to high speed ships. Has, however, been found to be beneficial in relatively low speed ships such as tankers and bulk carriers. Research has shown that the bulb has a greater effect when situated some distance forward of the stem. This has led to adoption of the ram bulb.

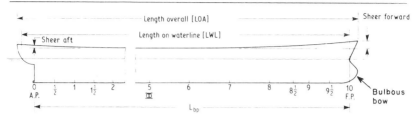

Figure 5. Terms used in naval architecture (ii)

Bulk carrier

Specialized 'dry cargo in bulk' carrier with large cargo hold volume. Well suited for cargoes such as coal, grain, bauxite and sugar. Single deckers with machinery aft as is also, in general, the accommodation. Can be loaded without having to trim cargo. This is attained by sloping plating which extends from hatch sides to ship's sides and from ship's side to tank top (Figure 6). This permits easy unloading of cargo as cargo will fall down to be below hatchway as unloading proceeds, allowing use of grabs.

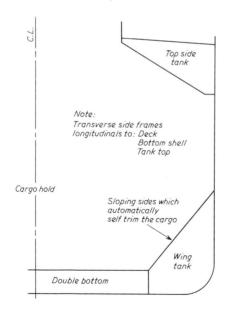

Figure 6. Bulk carrier, mid-ship section

Bulkhead	Vertical partition which subdivides interior of ship into compartments.
Bulwarks	Vertical plating erected at gunwales of ship to reduce quantity of water breaking into deck in a seaway.
Bunker fuel oil	Primary source of heat for ships and land plant. Residual fuel remaining after maximum amount of petroleum and light fuels have been distilled off. Contains leftover products of distillation; fuel quality will decrease and viscosity increase as fuel prices rise due to reduction in world oil reserves. Fuels with viscosities of 5–6,000 Redwood No. 1 seconds are being mentioned for the future and ships' boilers could be one of the few plants in which it is economic to burn these fuels which will produce serious atmospheric pollution unless exhaust gases are cleaned, an expensive process.

The various fuels used at sea today vary widely and may be blended. The following range of kinematic viscosities can be taken as an approximate guide:

(1) Marine diesel and gas oil fuels:
2–24 centistokes (cSt) at 40°C
31–110 redwood No. 1 (seconds) at 100°F (37.8°C);
(2) Heavy fuel oil (residual):
30–460 centistokes (cSt) at 50°C;
200–4,600 redwood No. 1 (seconds) at 100°F (37.8°C).

Buoyancy	Support given to a ship by the water in which the vessel floats. Ship must have sufficient buoyancy to support the loads it is intended to carry.

$$\text{Buoyancy per unit length} = \frac{A}{\text{density of fluid}}$$

Where A = immersed cross sectional area of the ship.

Busbars (feeder)	Non-insulated copper bars at back of main switchboard to which feeders from main alternator or generator main breakers are attached.

Butane (C_4H_{10}) — Paraffin hydrocarbon obtained from casing head gases in crude petroleum refining. Supplied in pressurized containers (bottle gas) for cooking and heating. Known as liquid petroleum gas (LPG). Formerly transported in pressurized tanks but now refrigerated and carried in special liquified gas ships.

Butt strap — Connecting metal strap, rivetted or welded, covering butt joint adding strength to joint.

Butt weld — Weld made between two plates placed side by side without overlapping.

Butterworth system — System for cleaning cargo fuel oil tanks with seawater heated to about 80°C. Water jets $(180\,lb/in^2)$ are directed to all parts of cargo tank to remove crude oil deposits remaining from previous cargo. Resultant oil/water mixture is then pumped to slop tank for separation.

Butyl rubber — Copolymer of isobutylene and isoprene exhibiting good chemical and ozone resistance and low permeability to gases. It is resistant in many phosphate-ester fluids but not to petroleum-based fluids.

By-pass — Pipe used to control and divert circulation of a fluid, gas, etc.

C

Cable — (1) Rope, or chain, used primarily for mooring a ship, e.g. as the anchor cable. Use of cables for mooring led to a cable length becoming a convenient unit of distance measurement, particularly in relation to manoeuvering of ships close to shore or harbour works. One cable is equivalent to one tenth of a nautical mile, that is, very nearly 200 yd. (2) Insulated, and sometimes also protected or armoured electrical conductor. Term is generally used to distinguish between a single or twin insulated wire and a conductor with a relative massive single or multiple core. (3) Term 'cable' also frequently used to describe a telegraphic message transmitted over long distances, e.g. between continents, which implied originally transmission by submarine cables.

Cable, chain See *Cable*

Cable, electric See *Cable*

Cable stopper Device used to secure cable when riding at anchor and thus reduce load on windlass.

Calcium chloride Chemical formula $CaCl_2$. Readily absorbs moisture from the atmosphere and so is a useful medium for drying gases.

Calorifier Apparatus for heating water in a tank, frequently taking the form of a coil of heated tubes immersed in the water.

Calorific value Quantity of heat released per kilogramme of fuel completely burned is termed the calorific value of the fuel.

Cam Projection on a revolving shaft, or collar mounted on a shaft, so shaped as to impart some desired linear motion to a follower which is kept in contact with it. Used in machines, valve gear and fuel pumps etc. for converting a rotary motion into any desired reciprocating motion. During each revolution of a cam it is necessary to take up the movement in a controlled manner. Cam design takes various forms. Harmonic cams are the mathematically simplest form of cam profile made up from simple arcs and tangents. Non-geometric cams are not very suitable for high speed operation.

Camber Amount of transverse curvature on a deck; also called 'round of beam'. See *Figure 2*.

Camshaft Shaft carrying the cam(s) which operate valves. Camshafts for automobile engines and industrial engines are mostly made in high-duty cast iron.

Cant frames Frame not square to centre-line at stern. These are not required with a transom stern as flat stern plating can be stiffened with vertical stiffeners.

Capacitor

Modern term for condenser. Device consisting of two electrodes separated by an insulant, which may be air, for introducing capacitance into an electric circuit, and which has the capability of storing electrical energy. Fitted across ignition make-and-break to produce a hot spark at the plug in I.C. engine.

Capstan

Revolving device on vertical axis used for heaving-in mooring lines or anchor cable.

Carbon dioxide

Inert gas which is present in the products of fuel combustion – Chemical formula CO_2. Also used as a refrigerant gas and for firefighting in particular electrical fires.

Cardan shaft or joint

A flexible mechanical arrangement for taking up limited misalignment between a diesel engine and a propeller shaft.

Cargo heating coil

Tanks for heavy oils, molasses or other viscous fluids are fitted with heating coils so that the fluid may be sufficiently liquified to run freely to pump suctions.

Cargo manifold

A pipe branching into two or more flanged open ends. Usually fitted at the ends of the deck crossover pipes on tankers.

Cargo plan

Plan which indicates position in ship of different cargoes.

Cargo port or door

Opening in ship's side for loading and unloading cargo.

Carry-over

(1) Transfer figures to column of higher value. (2) Postpone to a later time. (3) Keep over to next settling day on Stock Exchange. (4) Water passing from boiler into steam range and machinery, possibly causing water hammer.

Carving note

Form completed by owner of ship under construction. Gives details of tonnage, name, port of registry, etc.

Cathode

Negative electrode. See *Anode*.

Figure 7. Cavitation tunnel tests on 4-bladed propeller

Cathodic protection	An electrolytic system for the protection of a metal from corrosion by providing an electric current to neutralize the current naturally produced in the corrosion process. The current may be provided by sacrificial anodes (see *Anodes, sacrificial*) or be applied through incorrodable anodes in an impressed current system. The electric circuit is completed through the corroding medium to the metal to be protected which is the cathode.
Caulking	(1) Filling seams of wood planks with oakum. (2) Method of closing butts and seams of steel plating to ensure watertightness.
Caustic embrittlement	Embrittlement of steel which occurs when it is exused to highly alkaline conditions. Phenomenon may occur in boilers where crevices, such as between tubes and tube plate, allow a concentration of caustic conditions to accumulate.
Cavitation	Is the formation of cavities round a pump rotor or propeller blade – often on the back of the blade – these cavities being filled with air or water vapour. The effect of cavitation on a propeller is twofold: (1) the cavities formed eventually collapse resulting in a severe mechanical action which produces erosion of the blade surface, (2) with severe cavitation there is a loss in propulsive efficiency. Experiments have been devised to study the pressure distribution round the blade by operating the propeller in a closed channel or cavitation tunnel in which the fluid can be controlled. (Figure 7).
Cement	Formerly used for protection of plating against the action of bilge water, before epoxy paints were developed, to give effective protection to steelwork.
Centre girder	Continuous girder in double bottom that runs fore and aft on centre-line.
Centre of Buoyancy (B)	Centroid of underwater volume of ship and point through which total force of buoyancy can be assumed to act. (Figure 8). An

27

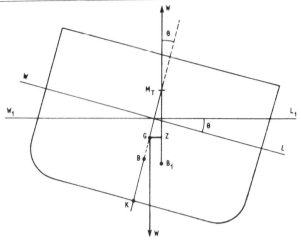

Figure 8. Transverse metacentre

important feature of form is the longitudinal distribution of displacement as expressed by the longitudinal position of the centre of buoyancy (LCB). For a ship to float on even keel, LCB must be under the centre of gravity and, to avoid excessive trim, position of LCB is dictated by the loading of the ship.

Centre of flotation (CF)

Centroid of waterplane area. For small angles of trim consecutive waterlines pass through CF.

Centre of force

Point at which total force exerted can be considered to act.

Centre of gravity (G)

Point through which total mass of ship may be assumed to act. See *Figure 8*. Position of centre of gravity of ship depends only on distribution of masses in ship. Position can be calculated but vertical position can be determined experimentally. See *Inclination test*.

Centre of pressure

Point at which entire pressure on an immersed area can be considered as acting.

Certificates of Competency

In addition to surveys required by classification societies, British ships have to undergo statutory surveys controlled by the

Department of Trade. For a passenger ship a Passenger Certificate can only be issued if the ship complies with the Merchant Shipping Acts. A Safety Certificate is also required to meet the provisions of the Convention 'Safety of Life at Sea' (SOLAS).

Cetane number

Number indicates ability of diesel engine fuel to ignite quickly after being injected into cylinder.

Chain locker

Compartment in ship into which anchor chain is hauled and stowed. Chain is stowed on a grating to permit drainage overboard through scuppers.

Chamber

(1) Enclosed space in a machine. (2) Compartment in a structure. (3) Apparatus designed to reveal tracks of ionizing particles by a visible effect – bubble, spark, etc.

Chamber of Shipping of U.K.

Body formed to promote and protect the interests of British shipowners. Received Royal Charter in 1922. Now known as the General Council of British Shipping.

Charpy test

Beam impact test in which a notched specimen is fixed at both ends and struck in the middle opposite the notch by a pendulum. The energy thus required to fracture the specimen is a measure of the resistance of the material to shock loads.

Charterer

Person or firm which engages the ship and with whom the shipowner enters into a contract. Any charterer of a ship, except a charterer by demise.

Chart room

Room all charts are kept in as well as other navigating equipment such as sextants; situated on navigating bridge/wheelhouse.

Chartered Engineer (C.Eng.)

Qualification awarded by the Council of Engineering Institutions to qualified engineers who normally have an engineering degree backed up with practical experience in a responsible position. See *Professional engineer* for a detailed definition.

Check valve

Non return valves which regulate flow of fluids to boilers, in piping systems and machinery.

Chemical carriers

Vessels specially designed to carry chemicals many of which are highly corrosive, poisonous and volatile. The type is related to cargo containment and survival capability.

Chlorine injection

A sea water circulating system becomes fouled by organisms unless a biocide is injected into it. Chlorine for this purpose can be obtained from a solution of sodium hypochlorite injected or from the electrolysis of sea water in a special apparatus.

Chock

Smooth surfaced fitting at weather deck side and through which mooring ropes are led. A wedge for securing a hatch cover or adjusting the alignment of an engine or gearbox.

Choke

A coil of high inductance. When used as a filter it impedes the current in a circuit over a specified frequency range while allowing relatively free passage of current at lower frequencies.

Chuck

(1) Part of lathe that holds rotating workpiece or tool. (2) Name sometimes given to fairlead.

Circuit breaker

Mechanical switching device capable of making, carrying and breaking currents under normal circuit conditions and also making, carrying for a specified time, and breaking currents under specified abnormal circuit conditions, such as, short circuit. A safety device.

Circulating pump

Centrifugal or axial flow type pump that draws water from sea to supply cooling water to condensers and also to machinery.

Cladding

Coating of one metal with another. True cladding involves rolling a thin layer of metal on to a basis metal or depositing a layer by welding.

Clarifier

Clarification is one of the operations in mechanical separation. The major applications are in cement, metallurgical and chemical industries, water purification and sewage treatment. The methods include filtration, sedimentation and centrifugation.

Classification society

Ships are 'classed' for seaworthiness by an independent Classification Society by means of periodic surveys. There are a number of such Societies such as Lloyds Register, Bureau Veritas, American Bureau, Norske Veritas, etc. Insurance of the ship is dependent on satisfactory classification.

Cleading

Any covering used to prevent the radiation or conduction of heat.

Cleat

(1) Fitting having two horns around which ropes may be made fast. (2) Clip on frames to hold cargo battens in place.

Clingage

Cargo remaining in tank of ship after discharge attached to sides, bottom, deckhead and girders of tank and is removed during tank cleaning. Term mainly applied to oil tankers

Clinker plating

Each strake of shell plating overlaps strake below.

Clogging indicator

Device operated by differential pressure across filter element, which indicates that element has reached its clogged condition.

Coagulator

This term is occasionally used to denote coalescer, but the latter is the preferred term.

Coalescence

Combination of oil droplets to form larger ones ultimately a continuous oil phase.

Coalescer

Unit containing material, the surface properties of which promote coalescence, i.e. the combination of small oil droplets to form larger ones which can then separate more readily under gravity and ultimately form a continuous oil phase. The coalescer is placed downstream of the gravity separator coalescing and separating those droplets which pass through the gravity stage as a result of their small size. A coalescer has in theory an infinite lifetime.

Coaming

Vertical plating bounding a hatchway. Heights of coamings depend on hatch position, some being more exposed than others. The coaming may be omitted altogether – flush coaming – if directly secured steel covers are fitted and the safety of the ship is ensured.

Coatings, protective

A covering adherent to a metal for the purpose of protection against corrosion. The term is usually applied to paints and similar compositions of organic origin.

Cofferdam

Narrow space between bulkheads or floors to prevent leakage between adjoining compartments.

Coffin plate

After plate of keel that connects with sole of stern frame.

Collet

Ring or collar, usually split, to encircle a groove in a stem or shaft where it is retained by an outer ring or seating. Used to secure the springs to the valves in the cylinder head of a car.

Collision bulkhead

Foremost transverse watertight bulkhead and extends to freeboard deck. Limits entry of water in event of bow collision damage.

Colza oil

Made from crushed rape-seed and used as a lubricant as well as an illuminant.

Combustion

Act of burning. Chemical action accompanied by heat.

Combustion chamber

(1) Space bounded by piston crown, cylinder cover or upper piston crown and cylinder wall in direct injection diesel engine, in which combustion takes place at or near top dead centre. (2) Chamber formed in cylinder cover of indirect injection engine connected to main cylinder by relatively narrow passage to promote turbulence. (3) Space adjacent to burner in which combustion takes place in boiler.

In compression ignition engines there are several ways of ensuring that the fuel shall be intimately mixed with air. The open or direct injection chamber relies on several jets to distribute the fuel in the air. In the swirl chamber air may be compressed into a separate chamber such that a rapid circulation past the injector may be obtained. With the pre-combustion chamber air is compressed into a separate chamber through a number of small holes. In spark ignition engines the ideal chamber is spherical with the spark plug at the centre.

Commutator

Mounted on the moving element of a rotating machine, it consists of a cylindrical ring or disc assembly of conducting members, individually insulated in a supporting structure, with an exposed surface for contact with current collecting brushes.

Companion

Permanent covering to a ladderway.

Companionway

(1) Set of steps leading from one deck or another. (2) The ladder used for disembarkation.

Compartment

Subdivision of hull by transverse watertight bulkheads creates compartments such that the vessel may remain afloat after flooding.

Compass adjuster

The magnetic compass is adjusted by a specialist known as a compass adjuster who carries out his work by a process called 'swinging the ship'. This is essential since true north and magnetic north are not the same.

Compass adjustment

The deviations of an installed magnetic compass are ascertained for different directions of a ship's head and are reduced as much as possible by fixing small magnets in suitable positions in the compass stand, so as to neutralize the magnetic effects of the ship and its equipment on the compass needles. The deviations remaining after the compass adjuster has completed his task are tabulated, or plotted, on a card called the deviation table. This may have three columns headed respectively: Ship's Head by Compass; Deviation; Ships Head Magnetic; and the corrections shown are applied in all subsequent navigational operations.

Compound wound

Applied to d.c. rotating machines, denotes that the excitation is supplied by two types of windings, shunt and series. When the electromagnetic effects of both windings are in the same direction, it is termed cumulative compound; when opposed, differential compound.

Compression ignition

Initiation of combustion of fuel in diesel engine due to high temperature and pressure of air compressed by piston movement towards top dead centre.

Compressor

Machinery used to increase the pressure of a gas. It may be reciprocating or rotary.

Condenser

(1) Chamber in which exhaust steam is led to condense into water. (2) Electrical condenser. See *Capacitor*.

Condenser, scoop

See *Scoop condenser*

Condition Monitoring (C.M.)

Method of determining when maintenance should be done instead of carrying out maintenance on an empirical calendar basis. Avoids unnecessary maintenance and stripping of machinery. Techniques involve visual examinations, performance checks (r.p.m., output and heat transfer), pressure tests, ultrasonic tests for thickness, insulation tests and vibration measurements.

Since 1971 opening of turbines at first periodical survey can be dispensed with if C.M. readings are satisfactory. C.M. may become an essential maintenance tool as the quality of fuel oils deteriorate due to world energy shortages and the chance of a ship being topped up with an incompatible fuel increases. Two good quality fuels when mixed can turn out to be incompatible and there is a lack of knowledge on this subject.

Conning	Directing course of ship.
Conning position	Place on bridge with commanding view used by navigators when conning vessel underway.
Constant tension winch	Towing winch designed to take predetermined load and if this is suddenly exceeded winch pays out wire; when excessive load is relaxed winch takes in wire.
Contactor	Mechanical switching device having only one position of rest, operating otherwise than by hand, capable of making, carrying and breaking currents under normal circuit conditions, including operating overload conditions.
Container	A portable compartment for the repeated carriage of cargo in bulk or package form. A standard cross-section of 2,435 mm × 2,435 mm and lengths of 6,055, 9,125 and 12,190 mm are used. Rigid and collapsible types are available.
Container ships	Containers are loaded at a factory, conveyed by road or rail to the dock side and then placed on board the ship. A cellular container ship carries containers in the holds and on the weather deck. In the holds there is a cellular structure of angle bars forming guides into which the containers are stowed one on top of another.
	These vessels are virtually single deckers with machinery towards the after end. Very large hatchways are enclosed by flush hatch covers and additional containers are stowed on open deck and anchored in position by wire ropes. Special cranes are required at container ports to cope with containers. These ships are faster than normal cargo ships and speeds of up to 26 kn are not uncommon.
Control	Any purposeful action on or in a system to meet specified objectives. Control may be open or closed loop, manual or automatic.
Coolers	Heat exchangers arranged to remove heat from a volume or stream of gas or liquid. Use may be

made of the latent heat of evaporation of another fluid, frequently water, to absorb the unwanted heat and improve the efficiency of the system.

Cooling system

A piping and heat exchanger network in which a fluid (usually water) is used to remove heat from a piece of machinery. The system normally includes a heat exchanger, which is sea water cooled but freshwater cooling is sometimes used to avoid corrosion problems.

Core plug

Plug for blanking off a core or fettling hole in an iron casting, e.g. access is required into the cooling water of a cylinder block during and after casting.

CO_2 Recorder

Instrument which provides a record of the carbon dioxide content in any enclosed space or in a gas stream such as engine or boiler exhaust.

Corresponding speed

Froude studied the wave pattern of geometrically similar forms at different speeds and found that the wave patterns appeared to be identical when the models were run at speeds proportional to the square root of their lengths. The corresponding speed is the speed of the model (v) where:

$$\frac{v}{\sqrt{L}} = \frac{v}{\sqrt{e}}$$

v = speed of model e = length of model
v = speed of ship L = length of ship

The formula is, therefore, used to estimate ship speed from the results of tank tests on the model.

Corrosion

Deterioration of a metal by the natural process which reverts the metal to a state of lower free energy, such as an oxide.

Corrosion – clover leaf

Pattern of wear which can take place in the cylinder liner of a diesel engine. Increased wear takes place between the lubrication quills due to the lack of highly alkaline lubricating oil at this part of the liner. The amount of oil present is not enough to neutralize the corrosive products of fuel combustion, and a clover leaf pattern of wear appears. Corroded area varies inversely as the distance between control points.

Corrosion inhibition

Prevention of corrosion may take two forms: (a) cathodic protection (b) application of protective coatings and inhibitors.

Corrosion piece

In a salt water cooling system made of copper alloys, corrosion of the alloys is reduced by the introduction of a length of steel pipe or iron sandwich piece. The iron or steel acts as a sacrificial anode thereby protecting from corrosion the adjoining copper alloys. Similar arrangements are fitted in steel or galvanized steel salt water system. Corrosion pieces are fitted in pipe sections where renewal is easiest.

Corrosive wear

Conditions in which corrosion and wear occur simultaneously. The most usual incidence is in fretting corrosion, where mutual movement of surfaces in close contacts causes oxidation and increased wear between surfaces, such as in a push fit.

Corrosive wear in diesel engines

Attributed to sulphur content of the fuel which by an etching process reveals the pearlite and stellite (phosphide) structure of the iron. Different mechanisms may operate on different parts of the same liner to give different results.

Cotter Pin

Pin, usually either tapered or split, inserted through a shaft and engaging with holes or castellations in a nut or collar so as to prevent accidental turning, particularly slackening. As a method of locking nuts this application is now becoming less common because of the introduction of lockwashers of various types and adhesive chemical preparations such as 'Loctite'.

A cotter may also take the form of a rectangular section tapered key, such as those in use for many years to set up or adjust the halves of big end bearings, typically those of steam locomotives. A further form, usually of circular cross-section but with a tapering flat machined on one side, is used to locate, for example, a crank on a shaft and to enable torque to be transmitted, e.g. on the pedal cranks of bicycles.

Cracking Processes

Refers to the breaking of large molecules to form smaller molecules and thus create lighter compounds. It is a conversion process whereby lighter oils are produced from heavy oils. Main types of conversion processes are (a) thermal cracking (b) catalytic cracking (c) hydro-cracking.

Crankcase monitoring

Indicating or recording conditions inside a crankcase by means of a system of sensors and instruments showing the extent of smoke, the ranges of temperature or proportion of inflammable gas or liquid particles present and to warn of imminent danger.

Crawl

(1) In an induction electric motor, when it may run up to only one seventh of full speed, due to a seventh harmonic in the field form. (2) On paintwork, a defect consisting of the formation of wrinkles before drying.

Creep

A continuous and slow change in the deformation of a stressed material. It is exhibited by most metals at elevated temperatures at stresses well below the yield point at room temperature. Boiler superheater tubes very susceptible.

Critical path analysis

Method of planning a complex operation, or series of interdependent processes, so as to reveal the timing of each necessary stage in order to ensure completion of the whole operation within target and according to programme. The analysis highlights the long lead items enabling these to be progressed.

Critical speed

Critical speed of a shaft is influenced by the stiffness of its bearings and their supports. The fundamental critical speed of a shaft is equal to its natural frequency in lateral vibration.

Critical temperature

(1) The temperature at which magnetic

materials lose their magnetic properties (about 800° C for iron and steel), or at which some change occurs in a metal or alloy during heating or cooling. (2) The temperature above which a given gas cannot be liquified, which limits the number of gases suitable for use in refrigeration, e.g. CO_2 is inefficient in tropical waters.

Cross curves

Curves of righting levers (GZ) plotted on a base of displacement for constant angles of heel. See also *Curve of statical stability* and *Figure 8.*

Crosshead

Lower end of a piston rod. Carries top end of connecting rod.

Cross ties

Horizontal cross ties are introduced in wing tanks to stiffen tank side boundary bulkhead structure against transverse distortion under liquid pressure.

Cross trees

Thwartship members on mast to increase spread of shrouds.

Crude oil

Oils and petrols are hazardous if they are volatile, i.e. if at normal temperatures they emit a vapour that can accumulate until an explosive mixture with air is created. The vapour may be ignited at a temperature known as the flashpoint. For crude oil in its natural state, this temperature is quite high. Crude petroleum is a dark brown or black viscous fluid with relative density varying from 0.80 to over 1.0.

Cruiser stern

Rounded stern which is hydrodynamically efficient and improves water flow into and away from propeller.

Cryogen

Freezing mixture; something mixed with ice to form a freezing mixture.

Curve of statical stability

Curve of righting arms to a base of angle of inclination for fixed displacement is called a 'curve of statical stability'. Such a curve is readily obtained from a set of 'cross curves'. See *Figure 8.*

D

Damping

A ship has six degrees of freedom – heaving, swaying, surging, rolling, pitching and yawing. The first three are linear motions. Rolling is rotation about a longitudinal axis, pitching is rotation about a transverse axis, yawing is rotation about a vertical axis. It is necessary to damp these motions and many devices have been suggested for damping the rolling of ships such as passive water tanks and activated fins. Damping units are fitted to control systems to avoid surging and excessive oscillation in machinery systems; friction and viscosity have same effect.

Data logger

Used in centralized instrumentation for continuous monitoring of machinery. Now mainly superseded by an integral system.

Davits

Supports under which lifeboats are stored and launched.

Deadlight

Side light that does not open.

Deadrise

Athwartship rise of bottom shell plating from keel to the bilge. Also known as rise of floor. See *Figure 2*.

Deadweight (dwt)

Sometimes termed deadweight carrying capacity, is the difference between the light and loaded displacements of a ship. The deadweight (dwt) comprises the cargo, stores, ballast, fresh water, fuel oil, passengers, crew and their effects.

De-aeration

Removal of air, e.g. from boiler feed water. See *De-aerator*.

De-aerator

Vessel in which boiler feed water is heated under pressure to remove dissolved air and minimize boiler internal corrosion.

Decibel (dB)

Unit of sound pressure. Magnitude of sound in air is best described by the root mean square (r.m.s.) value of pressure fluctuations which is easy to measure. A simple logarithmic ratio

compresses the scale too much so the unit decibel (dB) is used. Sound pressure level (SPL) in decibels is $= \log P^2/Po^2$
where P is sound pressure and Po a reference pressure of $2 \times 10^{-5} N/m^2$. The human ear can detect sound intensities between 10^{-12} and 10 W/m^2 without pain. Response of ear to changes in intensity is more logarithmic than linear.

Deck head Underside of a deck.

Deck, main Uppermost continuous deck running the full length of the vessel.

Deep tank Many cargo ships today have one or more deep tanks situated amidships. They extend the full width of the ship and the top is level with a tween-deck. They serve the dual purpose of carrying liquid in bulk and ballast when the ship is only partially loaded. They can also be used for bulk or general cargo and have hatches which are oil-tight. A longitudinal division is fitted to reduce free-surface effect.

Deep well pump A centrifugal pump, usually driven by an integral electric motor, placed at the bottom of a deep bore hole, or at any low point in a system containing fluids, for raising the fluid.

Deformation Change in shape of a component under stress. If the amount of applied stress does not exceed the elastic limit of the material, the original dimensions will be restored when the stress is removed; this is elastic deformation. If the stress is above the elastic limit, the resulting deformation will be permanent; this is plastic deformation.

Dehumidifiers Substances or systems which remove moisture from the atmosphere. Chemicals such as calcium chloride and silica gel absorb water from the atmosphere.

Demineralizing plant Plant for removing the last traces of impurities from water. Ion exchange plants can produce water having only 0.1 parts per million of impurity.

Demise	Temporary transfer of a vessel to another party such that the owner ceases to have any control for period of charter.
Demurrage	(1) The delaying of a vessel by the charterer beyond the allocated time for cargo loading or discharge. (2) The rate of payment or sum payable to the shipowner for the detaining of his vessel.
Department of Trade	Government authorities are concerned with the safety of ships and the well-being of all who sail in them. In the U.K. the authority is the Department of Trade. This Department is empowered to draw up rules by virtue of Merchant Shipping Acts extending back more than a hundred years. These rules are compulsory on the shipowner. If a ship does not meet the standards prescribed by the Authority it would not be allowed to sail.
Depth moulded	Vertical distance at amidships from top of keel to top of deck beam at side or underside of deck plating at ship side. See *Figure 2*.
Derrick	Woodspar or steel tube used in association with winch for discharging and unloading cargo. The rapid loading and unloading of cargo from holds is an important factor in the economic efficiency of the ship. This has led to the development of other means such as side loading through doors in the side shell rather than through hatches. This in turn permits the use of fork lift trucks. The adoption of cranes on the ship is another means of achieving the same end. See *Figure 3*.
Derrick, heavy	Strong derrick for lifting heavy masses. Usually has special heel fitting with a socket secured to the deck.
Desalination	The removal of salt from sea water. Ships' plants are usually of the multi-stage evaporating and condensing types, producing fresh water and discharging concentrated brine. See *Demineralizing plant*.
Desuperheater	Heat exchanger in which the superheat or part

42

of the superheat temperature of the main steam from the boiler is removed before the steam is used for auxiliary purposes.

Detergent

A substance, usually in solution, for degreasing and cleaning. They are surface-active agents and unlike soaps do not leave a deposit on the cleaned article.

Detonation

Detonating explosives are distinguished from burning explosives by their high speed and mode of explosion termed detonation. Detonation is of course faster than burning and a cartridge of high explosive detonates at a high regular velocity. Detonation usually requires a confined space whereas in the open air the explosive only hums, e.g. cordite is difficult to light in the open air but explodes if hung in an empty cartridge case and detonated.

Deuterium

Isotope of hydrogen having double the mass. Deuterium oxide (D_2O) is heavy water, with density 1.105.

Dew Point

Air can only hold a certain quantity of water vapour at a given temperature. The amount of water vapour present is important in relation to the maximum amount the air can contain at that temperature. The ratio is called relative humidity. If a sample of air is lowered in temperature then ultimately the air becomes saturated and further reduction creates condensation. The temperature when this happens is the dew point for that sample.

Diaphragms

(1) Partitions in instruments; (2) Disc pierced with circular holes. A diaphragm meter is an instrument with a diaphragm or plate inserted in the pipe and a hole in the plate permits the water to pass. The difference of pressure on each side of the plate gives a measure of the flow.

Digital

A digit is any number from 0 to 9. Digital computer is fed with numbers or words in the form of digits. Binary code commonly used in which the two digits 0 and 1 are put together to represent numbers. Instruments tend to be smaller and more accurate than analogue solution, e.g. pocket calculator versus slide rule; wrist watch versus digital watch.

Digital computer

Machine capable of performing arithmetic computation at high speed. Its capacity and speed of operation depend upon size and type of coding used. Digital computers are particularly suited to calculations of a repetitive nature. See *Digital*.

Din rating

(Deutche Industrie-Norm). German industrial standard similar to British standards.

Diode rectifier

Electronic valve containing an anode and a cathode which allows current to flow in only one direction, i.e. when the anode is positive. Also a semiconductor device having two terminals and exhibiting a non-linear voltage – current characteristic.

Direct Drive

The coupling of a propulsion engine to a propeller by a shaft without intermediate gearing.

Displacement

Mass of water displaced by ship.

Dissolved oxygen

Natural water, salt or fresh, contains a quantity of air which is expelled on boiling. The oxygen content is important in its contribution to the corrosion process. Sea water usually contains about 0.5% by weight of dissolved oxygen.

Dissolved solids

Impurities dissolved in pure water when describing the density of water in boilers and evaporators. The preferred unit is parts per million (p.p.m.), one part per million representing one part of solid matter dissolved in one million parts of pure water by mass.

Distillate fuel

Fuel extracted from crude oil by distillation. In marine practice the term normally refers to àny liquid fuel from gas oil to marine diesel oil in the viscosity range 30–50 Redwood No. 1 at 100°F (378°C).

Distillation

A process for converting a liquid into vapour, then condensing the vapour and collecting the liquid distillate. Fractional distillation allows mixtures of liquids with different boiling points to be separated. Method used to make fresh water at sea.

Distiller

Unit for converting sea or river water into distilled water for use as boiler feed and drinking water. Saturated or exhaust steam is circulated in a steam coil which heats the sea water under a vacuum, the resulting vapour being condensed in a cooler or distiller. Heating may be carried out by electric coils in the smaller units.

Distributed winding

Winding of a rotating electrical machine, the coils of which occupy several slots per pole.

Diversity factor

The ratio of the estimated consumption of a group of power-consuming appliances under normal working conditions to the sum of their nominal ratings.

Dock

A place where ships may be moored for loading, discharging, repairs or fitting out during constructions.

A dry or graving dock is pumped out and maintained dry while the ship is within.

A wet dock is usually a large area in which a ship remains afloat often isolated from tidal movements by a lock gate. A floating dock can be submerged sufficiently for a ship to be floated onto it and then raised to lift the ship clear of the water.

Dock dues

Payments made for use of dock and its equipment.

Docking plan

Gives essential information required by dockmaster. Consists of outboard profile and midship section. Frame spacing, extent of double bottom, decks, watertight bulkheads and machinery spaces are shown. Positions of all openings in shell below waterline, rise of floor, bilge radius, bilge keels, and bottom lingitudinals are indicated.

Dodger

Screen used as protection from spray.

Donkey boiler

Small, usually vertical, auxiliary boiler for

supplying steam to winches or deck machinery when the main boilers are not in steam. A donkeyman is a crew member operating such plant.

Donkeyman

Rating who attends a donkey boiler, and assists in the engine room.

Double acting

Steam engine or pump in which steam acts on both sides of piston.

Double beat valve

A balanced thrust control valve supplying steam to a steam engine. The valve has two plugs on the same spindle. The steam pressure acts on the top of one plug and the bottom of the other placing the valve in equilibrium, enabling the valve to be operated with minimum effort.

Double bottom

Space between outer and inner bottom plating of hull.

Double insulation

Denotes that, on an appliance with accessible metal parts, a protecting insulation is provided in addition to the normal functional insulation to protect against electric shock in case of breakdown of the functional insulation.

Double skin

The use of two separate material layers for containment or construction purposes. A double skin construction is used on the sides and bottom of container ships.

Dowel

(1) Cylindrical wooden plug in deck plank to plug bolthole. (2) Close fitting pin, peg, tube, or bolt for the accurate location of mating parts.

Drag line

When ship is launched into restricted waterway it is necessary to provide a means of arresting the motion of the ship after launching. One system is to use drag chains laid in the form of a horseshoe on either side of the ship with the rounded portion away from the water. This means that the forward portion of the drag is pulled through the remainder of the pile. The wire rope drag lines are attached to temporary pads on the side of the ship.

Drain cooler

A heat exchanger used to reduce the temperature of hot water drains before entering the feed tank. Normally incoming feed water to the boiler is used as the cooling medium thus increasing plant efficiency.

Draught

Depth of water at which a ship floats.

Simply the distance from the bottom of the ship to the waterline. If the waterline is parallel to the keel the ship is said to be 'on the keel'. If not parallel the ship is said to be trimmed.

If draught at after end is greater than at the fore end the ship is trimmed by the stern. If the converse applies the ship is trimmed by the bow or by the head. Draught marks are cut in the stern and stem which give the distance from the bottom of the ship and the figures are 10 cm high with a 10 cm spacing.

Dredgers

Dredging work demands dredger types that differ greatly in character and operation: multi-bucket, suction, grab, dipper and rock breakers.

Drilling rig

A structure erected over a well to carry the drill and machinery for boring the well.

Drill ship

Used for drilling for oil in water depth down to about 500 m. See also *Dynamic positioning*.

Dry liner

Thin walled tube pressed into bored-out engine cylinder.

Dry sump

Sump from which all oil collected is immediately scavenged and returned to separate tank.

Dual fuel engines

Engines designed to burn either oil or gas or a mixture of the two, with simple, frequently automatic, means of changing from one fuel to the other. Normally the gas is ignited by injection of 2–10% of full load oil but some engines have been designed to use spark ignition when running on gas.

Ductility

Ability of a material to withstand deformation without failure.

Duct keel

Space formed by twin longitudinal girders in double bottom and used to carry double bottom piping which is consequently accessible (Figure 9).

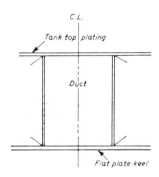

Figure 9. Duct keel

Dummy piston

Disc on shaft of reaction turbine to balance steam thrust on turbine blades.

Dump valve

Used in steam systems. It is a valve which may be opened to allow excess steam to be taken to a condenser in the event of a sudden reduction in steam demand. Fitted in plants with a high thermal inertia, such as nuclear reactors or exhaust gas boilers, where steam generated depends on engine speed rather than steam demand.

Duplex filter

An assembly of two filters in parallel with valving for selection of full flow through either filter.

Dynamic positioning

Highly sophisticated drilling ships today are generally fitted with fully automated dynamic positioning systems. A computer controlled system enables vessel to operate independently of anchors or any other mechanical mooring system. System can control bow thrusters, stern thrusters and main propulsion propellers.

Dynamics

Branch of mechanics that deals with the motion of bodies, matter under the influence of forces, or running machinery.

Dynamometer	Machine coupled to and absorbing the power of an engine on a test bed with means of measuring the engine power output.

E

Earth, electrical	Connection of any body to the main mass of the earth, by means of a low-impedance conductor, so as to maintain that body at earth potential. Connection with earth to complete circuit.
Earth lamps	Indicating lamps which, when connected between each phase or pole of an insulated system and earth, detect by comparative brightness the presence of earth faults on the system.
Earthed system	An a.c. 3-phase, 3 or 4 wire system in which the neutral or star point is permanently connected to earth. Also a d.c. system in which one pole is permanently connected to earth.
Easing gear	A means of manually operating safety valves usually from a position remote from the valves themselves.
Ebonite	A hard black material with good electrical insulating properties made by vulcanizing rubber. It contains carbon black and about 30% of sulphur.
Eccentric	Any circle revolving on an axis not in its centre. Misalignment.
Echo sounder	Electrically operated instrument that emits a sound from submerged surface of ship and measures time interval for return of echo. A scale converts interval to a depth indication.
Economizer	Used to transfer heat from boiler exit gases to feed water. Use of such a device increases efficiency of boiler plant.
Eddy currents	(1) Produced in any electrically conducting material when influenced by a coil in which

alternating current is flowing. Used in non-destructive testing for crack detection in castings and welds; not as penetrating as ultrasonics or radiography but have the advantages of speed, easily automated and no physical contact required with object under test. (2) Water particles moving past the hull in streamline flow. When the streamline flow freaks down the water particles revolve in eddies. The energy of this motion is wasted and can be treated as an increase in resistance.

Eductor

Another name for an ejector or jet pump.

e.h.p.

Effective horsepower, the power required to tow a ship without any appendages, i.e. without rudder, external shafting, 'A' brackets, bosses or bilge keels, in still air and smooth water.

Ejector

A type of pumping device used for discharging or expelling a liquid or gas from a space or tank. A jet of water, steam, or air is forced under pressure from a nozzle and creates a partial vacuum or low pressure area which acts on the suction pipe from the space to draw from it. No moving parts exist in this device.

Elastic limit

Highest stress which when applied to a material produces no measurable plastic deformation. See *Deformation*.

Elasticity, modulus

For a material worked within its elastic range strain $\propto$ stress, i.e.

$$\frac{\text{stress}}{\text{strain}} = \text{a constant.}$$

This constant is termed Young's modulus of elasticity and is denoted by E.

Elastomeric

The ability of a material after being stretched to return approximately to its original length. Elastomers include natural and synthetic rubbers and plastics with similar properties.

Electrode

Conductor through which electricity enters or leaves an electrolyte.

Electrolyte	A substance present in a solution in the form of electrically charged ions which is thereby capable of conducting electricity. Any liquid that can be decomposed electrically.
Electrolytic Action	Or Electrolytic Electrolysis action is caused when dissimilar metals are immersed in an electrolyte (such as salt water) and are connected together to form an electric circuit. This causes one of the metals to be attacked, and to be wasted away. Sacrificial zinc anodes are often used in order to prevent damage to underwater fittings from electrolytic action. They must be replaced as they gradually are eaten away.
Elongation	Extention of a material when a tensile stress is applied to it.
Emulsification	Mixing in very fine particles of two mutually insoluble fluids to form an intimate suspension of one in the other. Emulsification of water and oil sometimes produces viscous sludge.
End Float	Play or movement on a shaft in an axial direction; end play.
End tightened blading	See *blading — end tight*
Endothermic reaction	A chemical reaction which absorbs heat in the process.
Energy	The ability or capacity to do work. Solar energy is energy from the sun transmitted by electromagnetic radiation. Wind energy results from pressure variations in the atmosphere leading to the movement of large bodies of air. Wave energy is associated with wave height and results from the action of the wind on the sea surface.
Engine casing	Plating surrounding deck opening to engine room.
Entablature	Structure of a diesel engine above the bedplate and frames to which the cylinders are attached. In two-cycle engines the entablature is generally of box form and serves as a manifold supplying scavenge air to the cylinder.

51

Enthalpy

A measure of the total energy of a system including energy associated with the pressure–volume relationship plus the internal energy.

$$H = E + PV$$

where H = enthalpy or heat content
E = internal energy
P = pressure
V = volume.

The first law of thermodynamics states that the change in internal energy equals the heat absorbed less the work done by the system. The enthalpy of a substance is determined by its composition, temperature and pressure regardless of what has happened before and it is not therefore necessary to calculate absolute enthalpies.

Entrance

Immersed body forward of parallel body.

Entropy

A measure of the unavailable or waste energy in a closed system when mechanical work takes place; that is a system from which there can be no gain or loss of energy from its surroundings. The entropy change (ΔS) is equal to the heat (q) that would be expended during a system change with a small temperature alteration divided by the system temperature (T) in Kelvin or absolute units.

$$\Delta S = q/T$$

Entropy increases during an irreversible process such as when a solid changes into a liquid, a liquid into a gas, when hot and cold gases are mixed and during chemical changes. In these examples there is a decrease in the orderly array of constituent atoms resulting in an increase in entropy.

An increase in entropy is, therefore, associated with an increase in disorder. For example, when a solid changes into a liquid there is an increase in entropy as the molecules have greater freedom of movement (or more disorder) in the liquid than the solid state. In some crystalline solids the disorder is nil at an absolute temperature of zero; the entropy is, therefore, zero.

Epoxy resins

Chemicals produced from petroleum and natural gas are the bases of epoxy resins. These paints have very good adhesion.

Erosion

Wearing away of a material by the abrasion of fluids.

Evaporation

To turn a liquid into a vapour. See *Latent heat*.

Evaporators

Cylinders containing coils. Sea water is admitted to a cylinder and steam passed through coils causing water to be evaporated. The water vapour is then condensed and used for boiler feed.

Excitation winding

Winding of a rotating electrical machine for the production of a magnetic field.

Exhaust gases

Exit of gaseous products of combustion from a cylinder after completion of power stroke by piston, a gas turbine, or a boiler. Major source of waste heat.

Exothermic reaction

A chemical reaction which produces heat in the process.

Expansion valve

(1) An auxiliary valve fitted on some reciprocating steam engines used to provide an independent control of the point of cut-off. (2) A valve in refrigeration and air conditioning systems used to regulate the amount of refrigerant flowing around the circuit. When the liquid from the condenser is reduced in pressure by the valve, some liquid vaporizes cooling the rest down. This mixture, mainly liquid, then passes to the evaporator as the cooling medium. The amount of refrigerant passing through the valve is automatically controlled by the conditions at the evaporator outlet.

Extraction pump

This pump draws the condensate directly from the condenser of a steam plant and pumps it to the deaerator, usually against the considerable static head of the deaerator.

Extreme-pressure lubricant (EP)

Normal mineral oil containing added organic sulphur, chlorine and sulphur compounds which react at hot spots (above 200°C) between asperities to form films of sulphide, chloride and phosphide. These films shear more easily than substrata metal. Used on hypoid and other heavily loaded gears, also when speeds are insufficient to build up a thick film of lubricant.

Eyebolt   Bolt with an eye at end for hook.

F

Factories Act   Act under which 'Dock Regulations' derive their authority and which lays down regulations on how a factory should be organized such as safety, first aid and overcrowding.

Fairlead   Fitting that ensures a rope leading in a desired direction.

Falls   Rope used with blocks for lowering lifeboats etc.

Fans   A fan is fundamental to any system of mechanical ventilation. There are three main types. (1) Centrifugal – widely fitted as they give the best combination of high volume, high pressure and reasonably high efficiency. There are four types of impeller: forward curved (multi blade); backward curved (single blade); backward curved (aerofoil blade); and paddle blade. Efficiencies vary from 50% up to 85% for the aerofoil blade. (2) Axial flow – used less than centrifugal fans because of higher noise levels and lower pressure produced but fitted with silencer straight through design can be an advantage. (3) Propeller – limited, due to low pressure, to roof and wall extractors.

FAS   Free alongside ship. Shippers to arrange for delivery of goods within reach of ship's tackles unless custom point provides otherwise.

Fashion Plate   Side plate at end of superstructure deck which generally has an end that is curved.

Fatigue	Deterioration of the properties of a material which takes place under conditions involving fluctuating stress. Unlike brittle fracture, fatigue fracture occurs very slowly. Such fractures occur at low stresses which are applied to a structure repeatedly over a period of time. Often associated with sharp notches or discontinuities in structure.
Fatigue crack propagation	The progression of a crack in a material under a fluctuating stress cycle. The surface of a fatigue fracture sometimes exhibits couchoidal or wave marks which indicate the mode of fracture. Cracks may be arrested by drilling small diameter holes at each end of crack.
Fatigue test	A test by subjecting a specimen to a series of fluctuating stresses until fracture occurs. The stress may be applied in bending, torsion or axially in tension or compression (push-pull). The most commonly used machines (Wöhler) apply a weight to the end of a cantilever beam held in a rotating chuck, thus subjecting the specimen to a complete reversal of bending stress at each revolution; the number of applications of stress is recorded on a revolution counter. By subjecting a number of specimens of a material to fatigue tests with increasing loads, the minimum load at which a material will not fracture after a given number of stress reversals can be determined. This is known as the fatigue limit of the material.
Feed regulator	Controls the water level in a boiler.
Feed tank	Tank that feeds a service. Particularly a tank that contains feed water for boilers.
Feed water	Water supplied to boiler to compensate for water which has been vaporized.
Feedback	The transfer of a signal from the output of a system back to the input, to control or stabilize the circuit. A feedback control system is one where the signal is compared with the signal equivalent to the required system condition and corrective action taken if necessary. See *Hunting gear*.

Fender Resilient device, usually movable, interposed between a ship's hull and harbour walls or other ships so as to minimize impact and prevent direct contact so as to reduce risk of structural damage or chafing. Often of ropework, timber or pneumatic construction.

Ferrule Metal ring or cap, strengthening or forming a joint. Used to retain tubes in condensers and heat exchangers.

Field-coil Winding on polepiece of motor or generator.

Filter (1) Electrical. Network involving inductance or capitance or both, which freely passes energy at frequencies within one or more frequency bands and feebly passes energy at all other frequencies. (2) Fluid filter. It is essential to keep a working fluid free from contamination to ensure reliability of the equipment. This can be achieved by the use of filters. Solid particles in a circulating liquid can be removed by fine mesh strainers in addition to felt or paper filters. Magnetic filters or separators can remove particles of ferrous, bronze and bearing metals. (3) Air filter. Fitted to oil and internal combustion engines to avoid excessive cylinder wear due to abrasive matter entering the engine. These consist of layers of gauze, felt, paper, etc. (4) Oil pollution filter. Material which allows water to pass through but collects or separates the oil and dirt. Filters of the coalescer type may be used as the final polishing stage in the overall separation process, but because they retain oil and solid particles they have a limited life proportional to the amount of contamination.

Filtration Process of freeing liquids from suspended impurities.

Fineness (1) The ratio of the area of waterplane to area of its circumscribing rectangle. Varies from about 0.7 to 0.9. (2) Value of block coefficient gives guide as to whether ship form is full or fine.

Finger plate (1) Fastened to a door to prevent marks or damage. (2) Fitted to a machine to indicate the running position of the rotor or a shaft.

Flame impingement

Burning gas reaching a combustion chamber wall. Excessive flame impingement on a diesel engine piston may cause burning erosion of the piston material.

Flameproof

Type of protection for the safe use of electrical equipment in hazardous zones. The term is applied to a device, the enclosure or casing of which is so constructed that it will withstand without injury any explosion of the prescribed flammable gas that may occur within it under practical conditions of operation. It will prevent the transmission of flame which may ignite any flammable gas present in the surrounding atmosphere.

Flare

The outward sweep of the hull above the waterline from the vertical plane which promotes dryness and is associated with the fore end of the ship.

Flare stack

An isolated chimney or pipe at the end of which waste or unwanted gases are burnt off. Oil production platforms have special nozzles which are designed to prevent the flame being extinguished in high winds.

Flash point

Temperature at which vapour from oils and liquids may be ignited.

Flexible couplings

In some situations where shafts have to be coupled there are complications which necessitate the use of a flexible coupling; for example, temperature differences can produce lateral and axial misalignment. Vibration and shock loading can occur. The types vary from the ordinary flange coupling to bushing the bolt holes with rubber. The rubber disc type can accommodate severe misalignment. Some flexible couplings use springs to give the resilience required. Other designs involve hydraulic and electro-magnetic couplings.

Float control

Buoyant ball or cylinder operating valve or cock.

Floodable length

Length of vessel which may be flooded without sinking below the margin line (*Figure 10*).

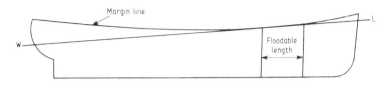

Figure 10. Floodable length

Floor ceiling

Wood covering placed over tank top for its protection.

Floors

Transverse vertical plates in double bottom. Provided both where bottom is transversely and longitudinally framed.

Flotation, centre of

See *Centre of flotation* (CF)

Flow meter

Used in systems for measuring fluid flow in closed pipes, conduits and ducts. The systems are broadly in two categories: inferential methods and volumetric methods. The former do not measure the volume of flow directly but infer it from velocity or change in pressure. The latter counts the number of times a known volume passes through the instrument.

Flue gases

Mixture of air and burnt and unburnt fuel leaving a boiler combustion chamber. The principal constituents are oxygen, nitrogen and carbon dioxide but some carbon monoxide may be present if insufficient air is available for combustion. Information about the combustion of fuel in boilers is obtained by the analysis of the flue gases. For general control work an apparatus is used in which 100 ml of gas is taken into a water-jacketed graduated burette and the constituents of the gas removed separately by absorption.

Fluidics

The use of mainly pneumatic devices for sensing, logic computation, and actuation. The process is based on the Coanda effect – the property of a jet stream to attach itself to an adjacent surface and remain there until some external force, normally another air jet, causes it to move.

Fluidised bed combustion

A form of combustion in which an inert material on the furnace bed is agitated by the combustion air and heated by directing a gas or oil flame on to the surface. On reaching the fuel ignition temperature, the fuel is admitted and causes a further increase in the bed temperature. Suitable for burning poor quality fuels. Has the advantage over conventional boiler of low combustion temperature and cleaner exhaust gas as some of the pollutants are retained in the bed.

Flush coaming

See *Coaming*

Flush deck

Upper deck without side to side erections.

Flying bridge

Uppermost deck-like platform.

Flywheel

The purpose of a flywheel is to create an available store of kinetic energy to suit the following purposes: (a) assist starting (b) reduce speed variation (c) prevent over-running if load is cut (d) limit phase variation. Flywheels are generally made in cast iron or steel. Rim speeds of cast-iron flywheels having an ultimate tensile strength of $230–290 \times 10^2$ N/cm^2 are:

Automobile petrol engines   70–78 m/s

Diesel engines   30–45 m/s

Foam

(1) Whitish froth that appears when water is agitated. (2) A suspension, often colloidal, of a gas in a liquid. (3) Chemically made foam sprayed on fire to extinguish flames by excluding oxygen (O_2) will provide limited cooling and reignition may occur if cooling water is not applied to reduce temperature. Specially safe and effective with petroleum and electric fires.

Foaming (lube oil)

Due to the churning action on the lubricating oil in pumps, gears or bearings, foaming may develop to an extent which interferes with the free flow through drain passages, filters or flowmeters, and this may seriously impair the effectiveness of the lubrication. In serious cases, the use of a lubricant containing anti-foaming additives may be the only satisfactory solution.

F.O.D.

Free of damage. Ship's officers report after each leg of a trip whether or not the ship has sustained any damage.

Forced draught

Air supply to a boiler furnace which increases the pressure and velocity of the combustion air by either maintaining the boiler room pressure positive in relation to atmosphere, or inducing draught in the uptake by blowing air through upwardly-directed jets. A positive air pressure may be obtained by means of fans, blowers, or by suitably shaped inlet ventilators making use of the local wind speed and direction.

Fore peak

Watertight compartment at the extreme forward end of ship.

Forefoot

Lower end of vessel's stem which curves to meet keel.

Foreign going ship

Ship trading to ports outside British Isles.

Forging

Deformation of a metal by applied force to change its shape and/or enhance its properties, such as by hammering or pressing.

Forming

To shape a beam, frame or other member to the exact form desired.

Fouling

(1) Restriction from movement by entanglement. (2) Marine growth on a ship which restricts its progress. (3) Deposition of substances within pipework which restrict the passage of fluid through it.

Frame

Transverse vertical member of hull structure that stiffens shell plating. Frames can also be longitudinal.

Free surface

When a tank is slack, i.e., contains a liquid but is not full, the stability of the ship is reduced by an amount depending solely on the extent of the free surface of the liquid and its relative density. The effect is independent of the amount of liquid of the tank.

Freeboard

Distance from the waterline to the upper surface of freeboard deck at side. (See *Figure 12*). Freeboard has considerable influence on seaworthiness of ship. The greater the freeboard the larger is the above water volume of the ship and this provides reserve buoyancy assisting the ship to remain afloat in the event of damage. Minimum freeboards are prescribed by International Law in the form of Load Line Regulations.

Freedom vessel

One of the many ship designs made available to replace Liberty class ships built during years 1942–5.

Freeing port

Opening in bulwark plating to free deck of water.

Free-piston engine

An engine in which the firing load on the piston crown is absorbed by compressing air or other gas below the piston underside which may be of larger diameter than the working piston. Free piston engines are usually of opposed piston design. The pistons are not constrained by connecting rods and crankshafts but there may be a mechanical linkage to keep the pistons in phase. Power output is in the form of compressed gas.

Not strictly a gas turbine but a composite engine involving a gas turbine as the power producer. The reciprocating portion serves as a gas generator giving no mechanical power but supplying hot compressed gas to the turbine.

Freon

Halogenated hydrocarbons (i.e. compounds derived from compounds of carbon and hydrogen by replacing some of the hydrogen by one or other of the halogens, chlorine or fluorine) used as refrigerants as they are virtually non-toxic and non-inflammable. Gases are expensive, odourless, non-corrosive and soluble in oil.

Freon 11 CCl_3F

Freon 12 CCl_2F_2

Freon 22 $CHClF_2$

Frequency changer

Device for converting a.c. electrical power from one frequency to another frequency.

Frequency filter

Network of resistance, inductance and capacitance to give minimum resistance to current over a designed frequency range and as much resistance as possible outside this range.

Frequency (Hertz – Hz)

A simple pendulum gives a type of motion described as oscillatory or vibratory. The maximum displacement is called the 'amplitude' of the motion and the time interval between two successive positions of the body having the same displacement is called the 'period'. The period of one complete vibration or oscillation could be very small and it is more convenient to consider the number of vibrations in unit time. This is referrred to as the 'frequency' of vibration. So that if T is the period and n the frequency then:—

$$n = \frac{1}{T}$$

1 Hz = 1 cycle per second.

Fresh water allowance (FWA)

Amount that load line mark may be submerged when loading in water of less density than that of sea water.

Fresh water generator

A piece of machinery used for converting impure water (seawater) to distilled or fresh water. The sea water, normally under vacuum, is brought ʋo the boiling temperature by steam, jacket cooling water, or electric heaters and the resulting vapour condensed into distilled water.

Fretting

Mutual movement of two components in close contact causing deterioration of the fit between them.

Froude number

Dimensionless parameter used to indicate the influence of gravity on fluid motion and in the study of wave making resistance. Generally expressed as $F = v/(gd)^{-\frac{1}{2}}$
where F = Froude number
$\quad v$ = gravity wave
$\quad g$ = gravitational acceleration of fluid
$\quad d$ = depth of flow.
See *Froude, William*.

Froude's law of comparison   If two geometrically similar forms – two ships or a ship and its model – are run at corresponding speeds then their residuary resistances per unit of displacement are the same.

So that if Rr is the residuary resistance of the ship and r_r that of the model then

$$Rr = r_r \left(\frac{L}{l} \right)^3$$

where L = length of ship
l = length of model.

Froude, William   b. 1810. Engineer and naval architect who influenced ship design by developing a study of scale models propelled through water and applying them to full size ships. Invented the deep bilge keel to reduce ship's roll and a dynamometer for measuring the power developed by large engines. See *Froude number*.

Fuel, Compatible/non-compatible   Two fuels from different sources may combine readily to form a mixture of intermediate viscosity, or they may tend to separate, or form sludge. Those which combine readily are said to be compatible, the others non-compatible.

Fuel, diesel   Diesel engines use a wide range of fuels varying from light gas oils to medium residual oils. The choice of fuel is controlled by engine characteristics such as cylinder diameter, engine speed and combustion chamber wall temperature. In general the fuel consumed in diesels is distillate oil. The viscosity of diesel fuels is of some importance. Oils of low viscosity are relatively easily atomized and fuels that are too viscous can spray droplets which escape complete combustion. Viscosity is specified in the U.K. in terms of redwood No. 1 seconds.

Fuel economy   The efficient use of fuel. In a ship, this may take any form from choosing the best sea route to take advantage of favourable wind, current or tide to ensuring that all machinery is maintained at its optimum running conditions.

63

Fuel – gas oil

An oil intermediate between light distillates and heavy diesel oils with boiling point above 244°C (400°F). Used in diesel engines.

Fuel, High Viscosity (HVF.)

An imprecise term applied for fuels with a viscosity in excess of about 1,500 redwood No. 1.

Fuel injection

See *Injection*

Fuel, low sulphur

Most fuels contain several percent of sulphur (S) which combines during combustion in a diesel engine to form sulphur dioxide (SO_2). This combines with the water vapour produced by combustion to form sulphuric acid (H_2SO_4) and lube oils are made alkaline to counteract the acidity.

However in recent years a few cases of sudden and catastrophic wear have been reported in diesel propelled ships after bunkering in communist countries with fuels containing less than 1% sulphur. The reported cases have been difficult to investigate thoroughly and the reason for the excessive wear rate has not been proved but it is thought that the low sulphur content in the fuel may affect the combustion flame.

Fuel, non distillate

The residue fuel left after the higher fractions have been distilled off. Used when running in some new engines as the fuel acts as a very mild abrasive in the cylinders.

Fuel, residual

In the refining of crude oil, many fractions are produced. The fuel which remains subsequently is known as residual. Bottoms is the left-over product from distillation of petroleum and from other refinery operations. Physical and chemical properties vary widely depending on crude oil sources and method of refining. Viscosity of Bunker C fuel is normally between 175 and 200 SFS at 50°C. Residual oils contain inorganic matter and the ash is often rich in sodium compounds which can lead to deposits and corrosion in boilers.

Full flow cargo system

A method of cargo handling employed on tankers, which uses large sliding sluice valves fitted in the tank bulkheads. By opening the valves the tanks are progressively drained aft where the cargo pumps extract the oil for discharge ashore. No suction pipework is, therefore, required other than a small stripping line.

Funnel

Uptake through which smoke, combustion and exhaust gases are led to open air.

Funnel guy

Normally there are at least four guys or stays to give support to the funnel.

Fuse

(1) Wire in a main or branch electrical circuit designed to melt when overloaded and thus interrupt current. (2) An apparatus for exploding explosives.

Fuse base

The fixed part of a fuse provided with terminals for connection to the external circuit.

Fuse carrier

The movable part of a fuse designed to carry the fuse link.

Fuse link

The part of a fuse, including the fuse element, which requires replacement after a fuse has operated and before the fuse is put back into service.

G

Galvanic action

Action which occurs during electrolytic corrosion. The term usually implies the presence of dissimilar metals in contact and their effect on each other in a corrosive environment. In such a system the anode (q.v.) is corroded and the cathode is thereby protected from corrosion.

Galvanizing

Generic term for any of several techniques for applying thin coatings of zinc to iron or steel to protect from corrosion.

Garboard strake

Strake of bottom shell plating adjacent to the keel plate.

Gas detection

The monitoring of an area or compartment for

toxic or flammable gases. It is usually associated with a measurement, an acceptable display or the triggering of an alarm in the event of an unacceptable value being recorded or registered by the equipment. See *Gas detection sensor*.

Gas detection sensor

Device whose resistance varies in the presence of gas. When forming one arm of a balanced circuit, an out-of-balance current resulting from change in resistance will indicate presence of gas. See *Gas detection*.

Gas freeing

Removal of pockets of gas from the tanks of an oil carrying vessel after cargo has been discharged. To avoid the possibility of explosive mixtures within the tanks an inert gas system is used. This is done by cleaning the tanks with detergents and ventilation.

Gas turbine

Rotary heat engine that converts some of the energy of fuel into work by using the combustion gas as the working medium. There are various types.

Gas welding

Process for joining metals by melting them with a gas flame from a torch. A concentrated gas flame was the first heat source for fusion welding. A variety of fuel gases combined with oxygen have been used to produce a high temperature flame, oxy-acetylene and oxy-propane being the most usual gases.

Gases

The following gases are described elsewhere in the glossary: Ammonia, Argon, Carbon dioxide, Helium, Hydrogen, Methyl chloride, Nitrogen, Oxygen.

Gasket

Material used to make a joint or seal between two surfaces, e.g. may be placed between two pipe flanges, or between cylinder head cover and cylinder.

Gauge glass

Glass tube or an arrangement of glass plates fitted to a gauge and used to give a visual indication of the level of liquid in a tank, pressure vessel, or boiler.

Gear generation

1. Individual cutters or abrasive wheels

These are generally employed in pairs, their inner surfaces being flat, unless modified in order to produce corrected teeth. They generate both sides of each tooth as it is rolled between them. Cutters may be used in a planeing or shaving sense, while if abrasive wheels are used the process is known as gear grinding, of which the 'Maag' system is typical and well known.

2. Rack or circular cutters

Straight-sided teeth cut on short lengths of straight rack, or circular gear-form cutters are used in gear shaper machines to generate involute profiles on spur gear blanks. The cutters reciprocate axially while being fed inwards slowly against the rotating blank, the gear-form cutters being themselves generated by straight rack-form tools. The 'Fellows' system of gear cutting is based on this principle.

3. Rotating hob

Cutters in the form of helical pinions have been further developed into the helical cutter with so small a lead angle that it becomes virtually a worm. The thread, or threads, of this are gashed and relieved to provide cutting edges at intervals along the length, and the cutting action of the hob results from its rotation in a form of milling machine. The 'Gleason' system is based on hobbing.

Gear grinding

Finishing the profile of gear teeth by grinding. Processes use shaped abrasive wheels which move relatively to the workpiece which is itself moved, to form the profile of the gear tooth.

Gear noise

Noise due to vibrations caused by impacts and non-uniform angular velocities of the rotating gears arising from pitch errors, inaccuracies of tooth profiles, eccentricity, etc.

Gear shaving

Improving the accuracy of the profile of gear teeth. The shaving cutter is similar to a helical gear in which the teeth have been serrated with cutting edges along the tooth spiral. The cutter is fed across the face width of the workpiece while in mesh with the teeth.

Gear tooth damage

(1) Abrasive wear: generally smooth and even wear resulting from prolonged service with moderately clean lubricant. (2) Scoring: a more fully developed and catastrophic form of scuffing, due to the same factors as scuffing, but may also be indiced by foreign matter or particles of debris in the lubricant. (3) Spalling: a pattern of very fine pitting, often so fine as to give an appearance as of the material being crystalline. It is due to impact forces and is generally to be found close to the pitch line. High speeds and pitch errors will tend to promote it. (4) Scuffing: surface damage of more severe nature than abrasive wear. Local welding and tearing of the surfaces is generally involved, and indicates inadequate lubrication and/or cooling having regard to the load and speed. Local overloading due to gear misalignment or incompatibility of materials may also be indicated.

Gear trains

1. Single reduction

Refers to a simple reduction gear train in which the reduction ratio is obtained in one stage by a pinion or pinions on one shaft meshing with a corresponding gear or gears on a single-output shaft.

2. Locked train

A form of double reduction gearing. A single input pinion driver. Two primary wheels via quill shafts on to two secondary pinions and finally to a single main wheel.

3. Split primary

Double reduction gear train in which the double helical secondary pinion is placed between each hand of the double helical primary wheel.

4. Split secondary

Double reduction gear train in which the double helical primary wheel is placed between each hand of the double helical secondary pinion.

General arrangement

(1) Plan showing general lay-out of ship design equipment in each compartment. (2) A drawing showing the component parts of an engine, or unit, in their correct assembled relationship. Such a drawing may consist of

three views, a plan and two elevations, and may include also one or more sectional arrangments showing internal construction.

General Average

General indemnity made by all interests concerned for a maritime loss deliberately but necessarily incurred for the safety of the remaining property when in peril. See *General Average Act*.

General Average Act

Voluntary and extra-ordinary action necessarily taken for preserving property in peril. The principle of general average is laid down in Rule A of the *York Antwerp Rules* 1950. 'Average' covers all damage sustained by ship or cargo during a voyage, as well as extraordinary expenses incurred in a maritime adventure. 'Average' also refers to partial loss.

General Average Expenditure

Extra-ordinary expenditure voluntarily and necessarily incurred for preserving property in peril.

General Average Loss

Loss that is due to a General Average Act or Expenditure.

General service pump

Pump that serves several purposes such as feeding donkey boiler, providing deck sea water service and feeding fresh water to drinking supply tanks. Cannot be connected to bilges.

Gimbals

Two rings, pivoted at right angles to each other and keeping a compass in the horizontal plane in all circumstances.

Girders

Continuous members running fore and aft under deck as a support.

Gland

Sleeve of soft material used to secure a tight packing on a piston, propeller shaft, pump spindle or electric cable. Synthetic rubber seals can be used.

Glass wool

Insulation material derived from molten glass. It is extremely light, vermin proof, fire-resistant, odourless and does not absorb moisture.

| Glow plug | (1) A heater installed in the combustion chamber of some diesel engines to assist in the starting of the engine from cold. The heater is switched off once the engine is running. (2) An igniter for re-lighting the fuel to a gas turbine in the event of the flame becoming unstable such as under cold conditions. |

GM (Metacentric height)

Distance from the metacentre (M) to the centre of gravity (G) of the ship. To be stable G must be below M. See *Figure 8*.

Goal post mast

Athwartship structure on weather deck to support derricks. Stump mast generally stepped in middle of structure.

Goose neck

(1) Swivel fitting on the end of a boom. (2) Tube turned over at head to prevent entry of water.

Governor

The centrifugal governor conical pendulum is one in which the rising of a rotating ball with increase of speed, controls the speed of an engine by operating levers to check the supply of fuel.

Graving or dry dock

Dock in which ships may be repaired or built, the water being pumped out as required.

Gravity separator

Chamber through which oily water is caused to flow in a quiescent manner allowing oil droplets to separate from water under the influence of gravity as a result of the density difference between oil and water. The term includes plate separators, a variant in which parallel plates are inserted to enhance performance. The bulk of the oil contamination is generally removed by a gravity separator.

Greenheart

One of the heaviest, strongest and most resistant of all timbers to decay. British Guana is the principal source of supply. Used for piles and underwater work due to its resistance to tonedo attack.

Grommet

Soft ring used under a nut to secure watertightness.

Gross register tonnage	(g.r.t.) A value resulting from the underdeck volume or tonnage measurement together with the tonnage of the tween decks and all enclosed spaces above deck. Certain deductions are made for areas which are exempt from measurement.
Gross tonnage	Tonnage is a measure of the internal capacity of a ship and consists of the gross and net tonnage which are computed independently. The gross is a function of the moulded volume of all enclosed spaces. The unit is a function of cubic metres. The word 'tons' no longer used.
GRP	Abbreviation for glass reinforced plastics. The glass in very fine filaments is mixed with a plastic before the latter is cured resulting in a finished product with enhanced properties.
Grubscrew	Setscrew without a head threaded full length with slot for turning, usually for securing a pulley or collar to a shaft.
Gudgeon	Boss on rudder post to take rudder pintles about which rudder turns.
Gudgeon pin	Short shaft connecting piston and connecting rod in a trunk piston engine, oscillating in bearings in one or both components known as 'piston pin' in America.
Guide and slipper	Two interacting components carrying connecting-rod side thrust in a crosshead engine. The slipper or crosshead shoe is attached to the crosshead and piston rod, and slides in the guide.
Gunwale	Junction of the upper deck with the sheel plating.
Gusset plate	Bracket plate connecting two members of ship's structure such as side frame and inner bottom plating.
Guy	(1) Rope used to guide and steady load while hoisting. (2) Rope or wire used as a stay.

71

Gypsy   Small drum attached to winch or windlass.

Gyro-compass   Gyroscope so mounted that the diurnal revolution of Earth is made to constrain North–South line of compass to seek the meridian and remain in it.

Gyroscope   Rapidly rotating wheel so mounted to have three degrees of freedom. Active stabilizing systems have gyroscopes as part of their control system. A massive gyroscope can be used to stabilize a ship but not used due to large size and mass.

H

Hamper   (1) Necessary but cumbrous equipment. (2) Top-hamper. Extensive superstructure above weather deck.

Hardening   (1) Increasing the hardness of a metal by heat treatment. (2) Work hardening: Increasing the hardness of a metal by deformation below the temperature at which softening occurs. (3) Case hardening. Pruducing a zone on the outside of a component which is harder than the core of the metal. This may be done by carburizing steel to increase the carbon content of the surface layer or by nitriding. Flame and induction hardening produce similar effects on certain steels.

Hardness, metals   Resistance to plastic deformation. The kinds of hardness considered for engineering purposes are: indentation hardness, wear hardness and scratch hardness. Tests are available to measure these properties of which the indentation test has the widest application. This test can be rapidly made without the destruction of the part involved; it is used in inspection to check the quality of products and their heat treatment. Hardness tests are used also to check strength and ductility.

Hardness, water   Due to the salts of calcium and magnesium. The insoluble carbonates cause deposits in boilers and combine with soap forming insoluble compounds resulting in inefficient laundry work. See also *Alkalinity*.

Harmonics   Components of a cyclical movement repeating two or more times per cycle.

Hatches   The openings in the decks of general cargo and bulk cargo ships to allow loading and discharge of cargo. Closure is by means of a hatch cover onto a coaming and must be watertight when fully fastened down. On ships such as tankers the tank hatch, which is much smaller, is only a means of entry for personnel to carry out inspection and maintenance. A hinged lid enables watertight closure when fully fastened down.

Hawse pipe   Tube through which the anchor cable is led.

Health monitoring   Normally applied to checks carried out to ensure that a machine or process will not endanger the health of the work force or the public, e.g. nuclear radiation, radiography, dangerous chemicals, etc. Also applied to condition and performance monitoring of machinery.

Heat balance   The equality of the heat released by fuel combustion in an engine to the sum of the useful work done, the heat lost to exhaust, coolant and lubricant, and the heat dissipated from the engine structure.

Heat balance calculation   Calculation of the quantities of heat discharged from an engine as useful work, heat to exhaust, heat to coolant and lubricant, and heat dissipated from the engine structure; and the equation of the sum of these to the heat available in the fuel consumed.

Header tank   Container connected to an engine cooling system, generally at the highest point, partly filled with water. The air or other gas above the water allows for expansion and contraction due to temperature change and may be vented to atmosphere or allowed to reach a higher pressure limited by a relief valve.

Heat exchanger

A unit for cooling engine lube oil and steam drains, etc., the cooling medium being sea, fresh or feedwater. To avoid the deposition of salt in marine engines, the direct sea water cooling temperature should not exceed 57°C.

A better procedure is to use a closed circuit with fresh water in association with a heat exchanger to transfer the heat to the sea water. By this method corrosion difficulties are avoided.

Heat transfer

Movement of heat from one place to another by conduction, convection or radiation. In engine practice the term is generally used for movement of heat from a fluid to a solid or vice versa as for example between combustion gas and piston or between cylinder liner and cooling water.

Heave to

Bring vessel to rest with head to wind.

Heaving

Vertical movement in a sea is called heaving. See *Swaying*.

Heavy fuels

The characteristics of fuel oils vary a great deal and the most important between the grades is in viscosity. The terms light, medium and heavy are applied to fuel oils of low, medium and high viscosity respectively. See also *Bunker oil* and entries under *Fuel*.

Heavy lift vessel

Vessel fitted with strong derricks and gear specially designed to lift heavy masses.

Heavy water

Water with a density about 10% greater than that of ordinary water. Used in heavy-water reactors.

Heel

Inclination of a ship from the upright (θ). See *Figure 8*.

Heel block

Block at lower end of derrick.

Heleshaw

The name of a type of hydraulic pump. It consists of a number of pistons arranged radically around a shaft. The pump runs at constant speed, variation of output being achieved by varying the path or length of stroke. The pistons must follow as they rotate around the shaft axis.

Helical gearing

Helical gears connect parallel shafts and the teeth wind helically round the axis. There are two types – single and double. The tooth action of the single helical gear produces end thrust. In the double helical gear the end thrust from one helix is balanced by that from the other helix. Helical gears are suited for very high peripheral speeds over 300 m/min. One used in reduction gearboxes.

Helical gears, double

Under load single-helical gears generate an axial thrust reaction which may be objectionable. Double-helical gears with the teeth cut on equal, but opposed helices, are often employed to cancel out the thrust forces. See *Helical gears – single*.

Helical gears, single

As with straight spur gears, these are used to connect parallel shafts, but the teeth are cut helically on the cylindrical blanks, in the form of a very coarse pitch, multi-start thread. This configuration provides a stronger tooth (considered as a cantilever) and generally smoother transfer of load from tooth to tooth, than with straight spur gears. See *Helical gears, double*.

Helium

An element; an inert gas present in the atmosphere. It is very light and used for filling balloons. It is also used as a shielding gas in helium arc welding. Chemical symbol–He.

Helix angle

The angle between the tangent at any point to the helix along which the tooth is cut and the plane through that point containing the axis of the gear.

Helm indicator

See Tell-tale.

Helmsman

Person who steers a vessel underway.

High-speed diesel

Diesel engine with a crankshaft rotational speed of 1,000 rev/min or more.

High sulphur fuels

Oil fuels containing more than 1% of sulphur.

Hogging	When a vessel drops at the extremities. The opposite is 'sagging'.
Holding down bolts	Bolts connecting a machine bedplate to a foundation such as tank top girders.
Homogeneous cargo	Entire cargo of the same uniform type.
Hooke's joint	Used in automotive transmission where angular displacements can occur. See *Universal joint*.
Hot spot	(1) An area of material on a rubbing component of a machine, as for instance a piston or bearing which has become excessively hot by malfunction such as a lubrication failure. (2) In carburettor engines an area of the inlet manifold wall heated by exhaust gas to aid fuel vaporization.
Hotwell	Chamber in which condensed exhaust steam from an engine is stored in an open feed system.
Hovercraft (or Air cushion vehicle (ACV))	The basic principle in such vessels is the generation of a cushion of air between their underside and the surface over which they travel. In this way relatively low powers can be used to propel the vessel at high speed. The system has been adopted successfully in cross-channel services.
HP Stage	The particular point in a process where a high pressure substance is utilized or produced. The high pressure stage of a steam turbine plant utilizes high pressure steam for power generation. The high pressure stage of an air compressor produces high pressure air.
Hull cleaning	Removal of fouling organisms from the external surface of the hull of a ship. This may be done in dry dock by brushes or in wet dock by divers or special equipment having rotary brushes.
Hull efficiency	The ratio effective power/thrust power. May be in the region of 98%.

Hull resonance

When a ship moves through regular waves a vertical force is created and heaving motion is thus generated. The magnitude of this motion depends on, among other factors, the ratio of the heaving period (T_H) of the ship to the period of encounter of the waves with the ship (T_E) Maximum amplitudes of heave will occur when T_H/T_E approaches unity. This is the resonance condition.

Hunting

Prolonged self-excited oscillation of any variable.

Hunting gear

A form of mechanical feedback. In the case of steering gears the position of the rudder stock is transmitted through mechanical linkage to the control rod of a variable delivery pump. As the rudder starts to move the hunting gear starts to remove the stroke from the pump so that the pump is off stroke when the rudder reaches the required position.

Hydraulic fluids

The three principle types are aqueor solution of glycols, synthetic fluids such as phosphate esters and water-in-mineral oil emulsions. All are highly fire-resistant but are flammable.

Hydraulic nut

See *Pilgrim nut.*

Hydraulic winch

A winch powered by an hydraulic motor. A central pumping station is often used to supply several winches and other hydraulic services on board ship. Advantages of the hydraulic drive include greater flexibility in positioning and installing, limited maximum motor output torque, variable speed control, good winching characteristics and reduced electrical loading.

Hydrazine (N_2H_4)

Colourless alkali N_2H_4 used as reducing agent, as rocket propellant, and for removal of traces of dissolved oxygen from boiler feed water.

Hydrodynamic

Deals with forces acting on or exerted by liquids, particularly water. In hydrodynamic lubrication the bearing surfaces are separated by a fluid of lubricant that prevents contact between the bearing surfaces. Hydrodynamic power transmission – there are three types:

(a) hydraulic coupling, (b) torque converter and (c) reaction coupling.

Hydrogen embrittlement

Hydrogen in the nascent or monatomic form is readily absorbed by most metals and in the case of steel it causes loss of ductility or embrittlement. Nascent hydrogen is formed when produced by chemical reaction such as when steel is pickled in acid. It is found in some boiler defects.

Hydrofoil

Frequently referred to as 'ships with legs'. A hydrofoil is a wing that operates in water and as the vessel moves off and accelerates the foil lifts the craft until the hull is clear of the water surface. Some of such craft are designed for speeds in excess of 100 kn.

Hydrogen

An element; a light gas which forms an explosive mixture with oxygen to form water. Chemical symbol – H.

Hydrogen burning (steel)

If steel is heated to a temperature above 700°C in the presence of steam, a reaction occurs between them which results in the rapid oxidation of the steel which may be described as burning. Some marine boilers have caught fire and melted due to this phenomena.

Hydrometer

Instrument to determine the relative density of liquids. The most widely used are the API (American Petroleum Institute) and the Baume. See also *Twaddle Hydrometer*.

Hydrostatic curves

The variations of ship hydrostatic data with draught are shown by a set of curves. Extremely useful in the assessment of end draughts and the stability of a ship in various conditions of loading. The calculations for such curves are now normally made by a computer.

Hygrometer

Instrument for measuring humidity of air or gas.

Hypoid gear

Form of bevel gear used in rear axle crown and bevel combinations in which the bevel pinion meshes with the crown-wheel below its centre line, giving a lower propeller shaft line.

I

Ice breaker

A ship specially strengthened and constructed to break up ice in order to open up a navigable channel for other ships to use. There are two main types: Polar and Baltic. The Polar is used for heavy duty operations.

Ignition

(1) Mechanism for starting combustion of mixture in cylinder of internal combustion engine. (2) Spontaneous ignition occurs in a diesel engine without the help of a spark when the compressed mixture reaches the ignition temperature of the fuel. (3) Magnetic ignition – there are two types: the rotating armature and the rotating magnet. (4) Coil ignition consists of an electric battery and coil which transforms battery voltage to a high voltage to produce a spark. (5) Ignition, CD is mainly fitted to outboard petrol engines. Each sparking plug has a separate coil and there is no distributor. Rotating magnets in the flywheel generate a voltage which is stored in a capacitor. A sensor magnet in the flywheel triggers the discharge of this through the appropriate coil, which multiplies the voltage to give the spark at the plug. CD ignition systems can eliminate the need for contact breakers and other moving parts, to give increased reliability. They generate much high voltages (up to 50,000 V) allowing the use of surface gap sparking plugs in which the spark can jump from the central electrode radially in any direction: thus greatly reduces fouling and eliminates the need to adjust the gap from time to time.

Ignition delay

Period in a diesel engine between commencement of injection and start of combustion.

i.h.p.

The indicated horsepower of the engine is the brake horsepower (b.h.p.) divided by the mechanical efficiency. So termed because it is usually obtained by calculation from the measured or estimated indicator diagram of the pressure cycle in the cylinder.

IMCO

Inter-Governmental Maritime Consultative

Organization. The United Nations Conference in 1948 created a new organization (IMCO) to provide co-operation among governments on technical matters affecting international merchant shipping.

Due to change their title in the 1980s to the 'International Maritime Organization', a specialist agency of the United Nations.

Impact test

Test to measure the resistance of a material to a suddenly applied or shock load. The most popular test is the Charpy in which a pendulum strikes a notched specimen and the amount of energy absorbed in breaking the specimen is measured.

Immersion (TPC)

Tonnes per centimetre. The mass to be added or deducted from a ship to change the mean draft by 1 cm.

$$\text{TPC} = A/97.5 \text{ in salt water}$$
$$= A/100 \text{ in fresh water}$$

where $A =$ water plane area is M^2.

Impedance

Property of circuit which determines the current produced by a known alternating voltage applied to the circuit, dependent on resistance, inductance, capacitance and frequency.

Impeller

The rotating component of a centrifugal pump or blower which imparts kinetic energy by centrifugal force to the fluid. Fluid enters at the shaft or eye of the impeller and passes via radial vanes out at the radius or perimeter.

Impressed current system

System of cathodic protection (q.v.) in which direct current is supplied through anodes usually of platinum to protect steel from corrosion in sea water. The steelwork forms the cathode of the electrolytic system.

Inclination test

Experiment to determine the vertical position of the centre of gravity of the ship for one specified ship condition. Test is carried out by moving masses across the deck under controlled conditions and noting the resulting angle of heel. Carried out on completion of construction and after major alterations affecting the stability.

Indicator diagram

Graph of work done by steam or combustion gasses while in a cylinder. Produced by mechanism in an Indicator.

Indicator gear

Apparatus for recording cylinder pressure in a reciprocating engine in relation to crank angle or piston stroke.

Induction motor

Motor in which rotor current is not drawn from the supply but is induced by relative motion of the rotor conductors; the rotating field being produced by stator currents when the stator is connected to the supply.

Inert gas system

To avoid the possibility of explosive mixtures in an oil tanker an inert gas system is installed to ensure that when the cargo is pumped from the tank it is replaced by non-explosive gas. Exhaust gas is taken from the funnel, cleaned by water washing, tested and pumped into the cargo tanks which remain pressurized when not full.

Infra-red

Energy radiation, below the visible spectrum, within the frequency band 10^7–10^9 kHz. Used for cooking in some types of galley equipment and in guided missile homing systems when the missile detects and locks onto ship and aircraft exhaust gases.

Injection

The process of introducing fuel into the cylinder of a petrol or diesel engine by means of a special pump and injection valve.

Injector

Device embodying a restricted nozzle through which fluid is passed at high velocity from a region of high pressure to one of lower pressure. Fuel injectors in diesel engines generally have a number of small nozzle holes and a spring loaded valve which is opened by fuel pressure. In steam plant, injectors passing cold water into a steam chamber are used to produce a vacuum.

Insulated system

An a.c., 3-phase, 3-wire system in which the

neutral or star point is not connected to earth. Also a d.c. system in which neither pole is connected to earth.

Insulation

(1) Any form of barrier restricting the flow of heat or electric current from one body to another is said to insulate them, and the non-conducting material used is termed insulation. (2) Materials of high resistivity and electric strength used to prevent conduction between a conductor and earth, other conductors or the frame of a device.

Insulation (refrigeration)

As the hull steel structure is an excellent conductor of heat, some form of insulation must be provided at the boundaries of refrigerated compartments. The main insulation materials used are cork, glass wool and polyurethane.

Insurance, insure

Make payment of money against loss or damage or injury to property or person. Policy: Signed contract of an insurer to make good a loss against which insurance has been affected. Broker: Person acting as intermediary between those requiring insurance and those willing to insure.

Integrated circuits

An integrated circuit is the combination of the equivalent of a number of discrete components, such as transistors, resistors and capacitors, into a single device resulting in miniaturization and increased reliability. The components are formed on a single chip of silicon and connected to form the required circuit by means of a metallic interconnection pattern which is deposited on the chip. The metal used is usually aluminium. These devices show increased reliability.

Inter-coastal

Calling at ports along the coast.

Intercooler

A cooler fitted between stages as in a multistage air compressor. The cooling medium is often sea water. The effect of intercooling in a compressor is to reduce the air temperature at constant pressure and thus reduce the amount of power required to compress the air in later stages.

Intercostal	Longitudinal girder between the floors and frames and thus not continuous.
Interface	The boundary between two surfaces in chemical contact.
Intrinsic safety	Method of protecting electrical circuits and apparatus for safe use in hazardous zones, in which any spark or thermal effect produced in normal operation or fault conditions could ignite the gas or vapour.
Inverter	Static device for converting electrical power in the form of direct current to electrical power in the form of alternating current.
Involute	The most usual type of curvature employed as the basic profile of modern gear teeth. Geometrically it is defined as the curve described by the end of a taut string as it is unwound from the surface of a circular cylinder, and the circle is referred to as the base circle. The true involute form is often modified by tip and root corrections to allow for tooth deflections under load and to promote smooth engagement and disengagement under operating conditions. It is a matter of convenience that the basic rack which generates involute teeth has straight-sided teeth itself.
Ionization	The process by which charged particles are formed from neutral atoms of molecules which substantially governs the electrical characteristics between electrodes in fluorescent or other gas filled tubes.
Iron	Ductile metal which with the addition of carbon becomes steel and with more carbon becomes cast iron with increased brittleness.
Isherwood type	Ship constructional system in which continuous longitudinal framing is the dominant feature.
Isolating valve	A valve in a piping system, normally fully open, which can be closed to separate one part of the system from another for use in an

emergency or during maintenance of part of the plant.

Isolator

Mechanical switching device which provides in the open position an isolating distance in accordance with specified requirements. It is capable of carrying current under normal circuit conditions and under abnormal conditions, such as those of short circuit, for a limited time.

Isothermal

Line on a chart connecting points representing the same temperature.

Isotope

One of two or more forms of an element differing from each other in atomic weight and in nuclear, but not chemical properties. The nuclei of isotopes contain identical number of protons, but different numbers of neutrons.

J

Jack stay

Taut ropes stretched for a specific purpose such as between heads of davits and between stanchions to take lacings of awnings.

Jack staff

Small staff erected at bow for a flag.

Jacket

(1) Enclosure surrounding a component such as an engine cylinder through which steam or water is passed to maintain a desired temperature. (2) Enclosure made of insulating material surrounding a component to reduce the transfer of heat or noise.

Jalousie

Slatted shutter.

Jerk note

Certificate given by Customs when ship has been searched and no unentered goods are on board.

Jerk pump

Mechanism used in diesel engines to produce high fuel pressure for a short period during each engine cycle to enable fuel to be injected into the cylinder.

Jet

(1) A stream of liquid projected forward or upward usually from a small orifice. (2) A

spout or nozzle through which a liquid is emitted.

Jib   Projecting arm of a crane.

Jig   Appliance that guides the tools operating upon it.

Jim crow   A look-out man, usually a trained observer. The lookout post may also be called the jim crow.

Job evaluation   A means of measuring the relative content and worth of a job. Usually achieved by an analysis of the job content under headings such as qualifications needed, specific skill requirements, environmental conditions, special hazards, etc.

Joggle   To offset a plate so as to avoid use of liners.

Joggle plate   Plate so shaped that longitudinal edge of plate curves and overlaps plate next to it.

Jointing   Material such as cork or asbestos based sheet placed between two hard surfaces to maintain a fluid-tight seal.

Joule (1 J = 1 Nm)   Quantity of heat. Work done when a force of 1 Newton (N) is exerted through a distance of 1 metre (m) in the direction of the force. Work done or heat energy expended by a current of one ampere flowing for one second against a resistance of 1 Ohm. Specific energy, calorific value of specific latent heat measured in joules per kilogram (J/kg) of the liquid or gas.

     1 Btu   = 1.055 kJ
     1 Btu/s = 1.055 kW
     1 Btu/lb = 2.326 kJ/kg

Journal   Part of shaft that rests on bearing.

Jumbo derrick   Derrick for heavy lifts.

Jumper stay   Stay going horizontally from one mast to next or other point; also called triatic stay.

Junction box   Joining place of electric cables.

Junk ring   Ring on upper part of piston.

Jury

(1) Temporary mast to replace one that has been damaged. (2) Mainshaft rig to replace rigging carried away. (3) Makeshift rudder constructed on ship when ship rudder lost or damaged.

K

'K' Factor

The load-stress factor for wear. The limiting load on gear teeth from consideration of wear depends upon the radii of curvature of the co-operating profiles, the pressure angle of the gears, the moduli of elasticity, the relative hardness and surface endurance limits of the materials. Typical values range from approximately 1,000 for 10×10^5 repetitions of stress to some 500 for 10×10^8 repetitions with hardened steel gears of 20 degrees pressure angle. For detailed definitions and information values, see Buckingham–'Manual of Gear Design, Section Two'.

Kapok

A mass of silky fibres that clothe the seeds of the ceiba tree. Used in life jackets, belts, sleeping bags and as an insulant.

Kedge anchor

One or more anchors carried in addition to the main, or bower, anchors and usually stowed aft. A kedge may be dropped while the ship is under way, or carried out in a suitable direction by a tender or ship's boat, to enable the ship to be winched off if aground, or swung into a

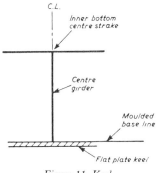

Figure 11. Keel

particular heading, or even to be held steady against a tidal or other stream.

Keel

Principal fore and aft structural member which runs along the middle of the ship's bottom. Often referred to as the backbone of the hull. (*Figure 11.*)

Keel blocks

Heavy wood or concrete blocks on which ship rests during construction and when in dry dock.

Keel plate

Strake of bottom plating on the middle line.

Keelson

Girder above bottom shell on each side of centre-line and running fore and aft.

Keep

Removable part of a housing holding a machine part in place, e.g. bearing keep.

Kent-ledge

Pig iron used as permanent ballast.

Ketones

The group contains low, medium and high boiling point solvents and also provides important intermediate materials for further chemical synthesis. Because of their high solvent power they are widely used in formulation of surface coatings and in extraction processes. One of a class of organic compounds containing the group CO with double bond to carbon of which acetone CH_3COCH_3 is one of the simplest.

Kerosene (paraffin)

Obtained by distillation of petroleum, coal or bituminous shale.

Key

Machined element used to connect a component to a shaft, e.g. a pulley or a flywheel may be keyed to a shaft. A key with a rectangular cross-section is called a flat key and is placed so that the smaller dimension is in the radial direction. Keys with a circular cross section are called pin keys and may be assembled along the shaft or through it.

Keyless propeller

See *Propeller, keyless.*

Keyway

Groove in which key fits. A keyway cut into a shaft weakens the shaft and introduces

87

D

concentration of stress. Consideration is given to this in the design stage.

Kinematic viscosity, (coefficient)

The ratio of the absolute viscosity of a liquid to its specific gravity at the temperature at which the specific gravity is measured. The coefficient of kinematic viscosity $(v) = \mu/\rho$ where μ = coefficient of absolute viscosity and ρ = density of fluid at temperature of liquid.

King post (or Samson post)

Vertical post fitted to support a derrick.

Kingston valve

A large valve fitted in a submarine's main ballast tank. The valve is used to control the flooding and blowing of ballast.

Kitchen rudder

Consists of two curved plates shrouding the propeller. For going ahead the two plates are parallel with the propeller race. For astern the plates are closed behind the propeller. Used on some small craft to obviate requirement for reversing gearbox.

Knee

Used to connect structural members perpendicular to each other such as deck beam and side frame.

Knot

Unit of speed. The international nautical mile is 1,852 m or 6,076.1 ft.
1 kn = 0.5144 m/s = 1 n. mile/h.

Knuckle

Abrupt change in direction of plating

Knurl

A protuberance. A knurled surface has a series of protuberances produced by milling, in order to facilitate grip by hand.

Kort nozzle

To increase thrust at low speeds a propeller may be enclosed in a nozzle. The patent Kort nozzle is often fitted to tugs and trawlers where under a heavy tow the propeller is working at a high slip.

L

Labyrinth-packing

(1) Seal between steam turbine rotor and casing. Series of fixed and rotating fins wire draw

turbine steam which builds up back pressure reducing steam to atmospheric pressure. (2) The efficiency of a sealing system controls the life of any bearing arrangement. Felt washers are a popular method of sealing and are efficient for low or moderate speeds, but are not suitable for oil lubrication. Leather and synthetic rubber seals are widely used on ball and roller bearing application. Various forms of labyrinth washers are extensively used for medium and high speeds and for both oil and grease lubrication.

Lamellar tearing

Stepped form of fracture in a series of tears which link up. It occurs in some rolled steel plate beneath welds and parallel to the plate surface due to contracting stresses which open up weaknesses in the through-thickness direction of the plate.

Laminar flow

If the water in the wake moves in a series of layers without mixing the flow is said to be laminar.

Lanyard

Rope or cord used for securing. Rope loop worn around the neck tied to boatswain's pipe. Part of a midshipman's uniform.

Lap

(1) The amount by which a slide valve or piston valve overlaps the steam and exhaust ports in a reciprocating engine. Steam and exhaust laps are necessary to provide for cut off before the beginning of the exhaust and to provide the cushioning necessary to reduce the load on the piston as it changes direction. (2) Joint in which one edge of a plate overlaps the other. (3) Fine form of hand grinding e.g. lapping a piston into a bore.

Lap winding

Distributed winding of a rotating electrical machine whose sequence of connections is such that it completes all its turns under one pair of main poles before proceeding to the next pair of main poles.

Laser

The word is a contraction of Light Amplification by Stimulated Emission of Radiation. A laser produces a concentrated unidirectional beam of monochromatic light. The active medium to produce the beam is contained in a transparent cylinder with a reflecting surface at one end and a partially reflecting surface at the other. The stimulated waves of light make repeated passages along the cylinder some of which emerge as a concentrated but narrow beam through the partially reflecting end.

Latent heat

The latent heat of fusion of a substance is the amount of heat required to convert unit mass of the substance from solid to liquid without change of temperature. The latent heat of vaporization is the amount of heat required to convert unit mass of a substance from liquid to vapour without change of temperature.

Lateral thruster

See *Bow thruster*.

Law of comparison–Froude

States that the wave-making resistance of a ship and its model vary as the ratio of the displacements or volume when the ship and the model are run at corresponding speeds. As for resistance experiments with models, the Law of Comparison may be used to predict the performance of geometrically similar propellers.

Layout

The drawing showing the positions of the items of plant and control equipment, in a ship's engine room or other space, and would normally be referred to as a plant layout drawing.

Lazaret

Storeroom or provision room.

Lead

(1) Lead. A lead weight attached to a rope for taking soundings in shallow water. The bottom of the sinker may be tallow or wax for taking samples of the sea bed. (2) Leads. Lengths of thick lead wire placed between journal and bearing cap during assembly for checking the bearing clearance. (3) Lead, slide valve. The amount that the steam port is open when the piston is at top dead centre and about to start its working stroke. Lead is employed to reduce wire drawing off the steam. See *Lap*.

(4) Symbol Pb. A soft heavy bluish-grey metal obtained mainly from the mineral galena. One of the most stable metals which is soluble in nitric acid but not in sulphuric or hydrochloric acids. (5) Amount by which one operation precedes another in a machine as for instance the number of crankshaft degrees in an opposed piston engine between the exhaust crank and scavenge crank. (6) Another name for an electric cable.

Lead screw

The master screw which controls the longitudinal motion of a lathe tool.

Lee

Area sheltered from the wind.

Liberty ships

Liberty ships were the product of a number of shipyards in the U.S.A. during the years 1942–1945. 2,700 Liberty ships were built on mass production lines.

Lid, valve

The component of a valve which closes the aperture of the valve.

Lifeboat

A boat carried by a ship for use in an emergency. It is provided with oars, a sail and sometimes a small engine. The lifeboat also carries water and provisions and is specially constructed to be seaworthy. Sometimes designed to protect the occupants from fire and foul weather.

Lift (valve)

The amount that the valve plug or lid moves away from the seat when the valve is opened.

Lighter

Open non-propelled barge.

Lightweight

The mass of the ship made up of the hull, propelling machinery and all fittings to make the vessel complete and ready for sea. Fuel and stores not included.

Lignite

Brown coal showing traces of plant structure. Intermediate between bituminous coal and peat. Used for domestic and industrial purposes, and manufacture of smokeless fuel.

Lignum vitae

Hard wood heavier than water. An evergreen tree of medium size with abruptly primate leaves, yielding a resin used in medicine. The heavy black heartwood is used for bowling balls, blocks, 'A' frame and stern tube bearings in ships which are water lubricated.

Limber hole

Drain hole.

Liner sleeve

(1) A separate and relatively thin sleeve fitted within an engine or pump cylinder to mate with the piston to form a renewable and more durable rubbing surface. A dry liner is in continuous contact with the cylinder wall and a wet liner is in contact on its inner surface with a cooling medium. (2) A sleeve or bush which is press fitted over a shaft in way of a bearing or gland seal. Any wear takes place on the sleeve which may be renewed, without damage to the shaft itself.

Lines of a ship

The ship's lines, usually drawn on a lines plan, define the shape of the hull up to the moulded dimension. Three basic views are drawn on the lines plan: the longitudinal elevation or profile showing the general outline, the half-breadth plan showing the shapes of the deck and water line, and the body plan giving the transverse sections. This information describes the hull shape in three dimensions.

Lines plan

Drawing that shows the form of a ship and consists of profile, half breadth and body plan. Shows three sets of sections through the form obtained by the intersection of three sets of mutually orthogonal planes with the outside surface. See *Figure 4*.

Lip seal

A means of leakage protection used on main propeller shafts to prevent the loss of lubricating oil or the ingress of sea water. It consists of a specially shaped stationary rubber ring held against a sleeve on the shaft by means of a spring and oil pressure, or by the water/oil differential pressure.

Liquefaction

Change of gas or solid into liquid state. Natural gas consists of methane, ethane, propane and butane. The propane and butane are removed by liquefaction and have a market as LPG.

Liquid natural gas (LNG)	Underground accumulations of gas of widely varying composition which may or may not be associated with accumulations of oil. Most oil accumulations have natural gas dissolved in the oil and often a 'gas cap' or free gas above the oil in the higher part of the structure or trap. Consists mainly of varying proportions of hydrocarbons in the paraffin series, the lightest member methane CH_4, boiling point 161.5°C, predominating (93–95%). Shipped in tankers at about 162°C in which state the liquid occupies one-six-hundredth of its gaseous state. Low temperature of liquid makes steel brittle and special materials/insulation required. Because of the low temperature cannot be distributed as liquid in containers.
Liquid slosh	The movement of liquid in a partially filled tank. The additional forces resulting must be considered in the design of tank bulkheads and sides. Alternatively wash bulkheads or reduced cross sections must be used in the tank construction.
Liquified petroleum gas (LPG)	Product from the production of gasoline which is undesirable in petrol engines because of its high volatility. The heavier gases propane C_3H_8 and butane C_4H_{10} are separated, compressed until liquid and stored under pressure in steel cylinders for industrial and domestic uses. Boiling points: propane 42.2°C, butane 0.5°C.
List	Transverse inclination of ship. Angle (θ). See *Figure 8*.
LOA	Length overall. Length of vessel taken over all extremities. See *Figure 5*.
Load line	The mark on a ship's side indicating the maximum draught to which a ship may be loaded under specified conditions in accordance with the Merchant Shipping Act 1932. Load Line Mark consists of a ring 300 mm in outside diameter and 25 mm wide intersected by a horizontal line 450 mm in length and 25 mm in breadth, the upper edge of which passes through the centre of the ring.

The centre of the ring is placed amidships and at a distance equal to the assigned summer freeboard below the upper edge of the deck line as shown in *Figure 12*.

The undernoted load lines are used:

Summer load line passes through the centre of the ring and marked S.

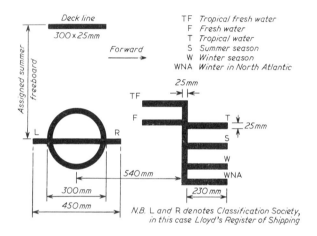

Figure 12. Load lines

Winter load line indicated by upper edge of line marked W and is obtained by adding to the summer freeboard one forty-eighth of the summer draught.

Winter North Atlantic load line indicated by upper edge of line marked WNA. For ships not more than 100 m in length it is the winter freeboard plus 50 mm. For other ships the WNA freeboard is the winter freeboard.

Tropical load line indicated by upper edge of line marked T and is the summer freeboard less one forty-eigth of the summer draught.

Fresh Water load line indicated by upper edge of line marked F and is the summer freeboard less $\Delta/40\,T$ cm where Δ = displacement in sea water in tonnes at the summer load line and $T = t/cm$ immersion in sea water at the summer load water line.

Tropical Fresh Water load line indicated by the upper edge of line marked TF and is the fresh water freeboard less one forty-eighth of the summer draught.

There are special markings for timer freeboards and sailing ships.

The initials of the authority by whom the load lines are assigned are indicated alongside the load line ring. The ring, lines and letters are painted in white or yellow on a dark ground or in black on a light ground. They are also permanently marked on the sides of the ships. See *Plimsoll line*.

Load Line Convention

Rules governing freeboard are laid down by an International Load Line Convention. See *Load line*.

Load-on-top

The practice of loading a fresh cargo of oil on top of oil recovered after tank cleaning operations. System now widely used in tankers engaged in the crude oil trade. The main object is to collect and settle on board water and oil mixture resulting from ballasting and cleaning of tanks. The oil/water mixture can be pumped ashore at loading terminals which have special reception facilities and thus reduce risk of oil pollution at sea. If the mixture cannot be pumped ashore the new cargo can be loaded on top and pumped ashore at the discharge port. Effective control of this system requires quantitative monitoring of the ballast prior to discharge.

Locker

Compartment in which gear may be stowed. See also *Chain locker*.

Locking nut

Fitted to secure the propeller on the shaft. See *Pilgrim nut*.

Log

Apparatus for ascertaining ship's speed through water and/or distance run.

Log book

Book in which events connected with ship are entered.

Logistics

The planning and organization of supplies, stores, and accommodation required for the support of large personnel movements and expeditions.

Loll

When the metacentric height is negative – G above M – then the vessel will 'loll' over until the centres of buoyancy and gravity are in the same vertical line. At the angle of 'loll' the ship has a positive metacentric height. See *Figure 8*.

Longitudinal framing

Hull framing that runs fore and aft instead of transversely.

Loop scavenge

Piston engine scavenge system in which both inlet and exhaust ports are at the same end of the cylinder and the inlet ports are so shaped that the air flows up one side of the cylinder, across the cylinder cover and down the other side before reaching the exhaust ports.

Low speed diesel

The low speed, direct coupled, reversible cross-head type of diesel engine is still a favoured form of marine propulsion. A popular proposition is the two-stroke direct drive diesel using the hot exhaust gases for a waste heat boiler. Speed range $100-150$ rev/min.

LPG (C_4H_{10})

Liquified Petroleum Gas. The market for LPG – propane, butane, etc. – is now considerable. Formerly carried in tanks in fully pressurized condition. Now refrigeration plants have led to the creation of specialized liquified gas carriers. See *Butane* and *Liquified petroleum gas*.

LP Stage

The low pressure point in a steam engine, steam turbine, or air compressor.

Lubricating oil additives

The performance of straight mineral oils can be improved by the use of additives that either enhance existing properties or confer properties not inherent in the oil. At one time additives were regarded with suspicion as implying an inferior type of base oil but nowadays their use is fully accepted and almost every commercial lubricant contains one or more additive.

Additives are generally used in relatively small proportions, not more than 5% wt, some even in parts per million. Blending components such as fatty oils, used in greater proportions, e.g. 5–20% wt, and soaps used in the manufacture of greases are not regarded as additives.

Additives usually consist of specific chemicals and are customarily named in terms of their function, e.g. anti-oxidants, dispersants, viscosity index improvers, pour point dispersants, and extreme pressure, anti-foam, anti-rust, anti-wear and multifunctional additives.

Lubricator quill

Attachment in the cylinder wall of a reciprocating engine through which oil is passed to lubricate the piston and cylinder.

Luff

(1) Weather side of vessel; opposite to Lee. (2) The weather edge of a fore and aft sail. (3) To bring a sailing ship's head into the wind.

Lug

(1) Projection from a casting. (2) Extremity of a shackle.

M

Magnesium (Mg)

Light brilliant white metal element (sp. gr. 1.7) which when alloyed with aluminium and other metals has a high strength to weight ratio and is used in aero-engineering. It corrodes rapidly in sea water. Used as deoxidizer for copper, brass and nickel alloys. Magnesium ribbon burns in air giving brilliant white light.

Main circuit breaker

Circuit breaker installed at the main switchboard. Hence, one to which the operation of other switching devices may be subservient.

Main inlet

Sea water inlet for circulating or cooling water pumps. Normally refers only to the larger pumps in the system.

Mandrel

Accurately machined bar or rod used for test or centring purposes.

Manganese (Mn)

Metal element (sp. gr. 7.4) added to alloys to impart special properties. In steel it refines the grain structure and imparts toughness. In aluminium bronze it improves castability.

Manhole	Hole in tank top, etc. to provide access.
Manifest	A complete list and description of the cargo which has been loaded into a ship and is on board. The list includes the numbers and marks on all packages, descriptions, weights and the names of the shipper and consignee.
Manoeuvring valve	A valve used to vary the amount of steam supplied to the main engine to alter the speed. Normal arrangement is one valve for ahead steam and one valve for astern steam.
Manometer	Pressure gauge for gasses and vapours.
Margin line	Line beyond which a ship should not sink. This is a line drawn parallel and 76 mm below the upper surface of the bulkhead deck at side. See *Figure 10*.
Margin plate	Outer boundary of the double bottom.
Marinisation	Modifications to machinery or equipment which was not originally designed for marine use to make it suitable for such use.
Marlin spike	Tapered metal pin used in splicing a rope.
Marline	(1) To moor or tie. (2) Small tarred line. (3) To bind with a line such that each turn is an overhand knot used for seizing and as a covering for rope.
Marry	(1) To interlace the strands of two ropes preparatory to splicing. (2) To place ropes side by side to be hauled simultaneously as when lowering a boat.
Martensite	A micro-constituent of steel having a needle-shaped structure. It is formed when steel is very rapidly cooled from the hardening temperature and is responsible for the hardness of quenched steel.
Mast step	The foundation on which a mast is erected.
Mast table	Small platform generally attached to mast to support hinged heel bearings of derricks.

Maul   Heavy hammer of metal or wood.

MCT   See *Moment to change trim.*

Mean time between failures A term used to indicate the reliability of an
(MTBF)   engine or machine. A bathtub curve gives the
                                instantaneous failure rate over the life cycle of
                                the equipment. In most cases the cycle starts a
                                period of early life failures when the
                                equipment is new, the failure rate then reduces
                                for most of the life of the machine only to rise
                                again at the end of the cycle when the machine
                                is wearing out.
                                MTBF is the total operating time of the
                                identical units being analysed divided by the
                                number of failures in a given time, e.g. 100
                                identical units in various systems each
                                operating for 16,000 hours with a total of 80
                                defects.

$$\text{MTBF} = \frac{100 \times 16{,}000 \text{ hours}}{80 \text{ defects}}$$

                                = 20,000 hours per defect
                                Failure rate = 100 × 16,000 hours
                                (Reciprocal of M.T.B.F.)
                                = 0.05 failures per 1,000 hours.
                                See *Reliability.*

Medium Speed Diesel   This type of diesel engine is being installed in
                                an increasing number of tankers, cargo liners
                                and bulk carriers. It is less costly than the slow
                                speed diesel engine and it operates reliably on
                                heavy fuel without undue maintenance. Speed
                                range 300–1,000 rev/min.

Megger   Apparatus for measuring insulation resistance
                                by generating a high voltage.

Membrane tank   Contains the LNG within a thin metallic
                                liquid-tight lining completely supported by a
                                load bearing insulation. This in turn is
                                supported by the structure of the ship.

Membrane wall   (1) A form of tube arrangement in a boiler
                                furnace. A strip of steel is welded between the
                                tubes making the furnace completely gas tight.
                                As the furnace is completely water cooled no
                                refractory is required in this part of the boiler.

(2) Double skinned tanks for the carriage in ships of LNG and LPG at very low temperatures, designed to allow for large expansion and contraction movements.

M.e.p.

Mean effective pressure. The constant pressure which, acting on the piston throughout each firing stroke of a reciprocating engine, would produce the same power output as the engine.

$$\text{m.e.p.} = \frac{\text{horsepower} \times 33.000}{\text{bore}^2 \times \text{stroke} \times \text{rev/min.}}$$

Mercury switch

Switch consisting of fixed contacts suspended above a pool of mercury, all being enclosed in a gas-tight envelope. Operation is responsive to mechanical movement and will occur when the mercury level is disturbed to form a conducting path between contacts.

Mesh

The travel of a tooth between engaging with and disengaging from the co-operating tooth is termed the meshing cycle. Gears may also be said to be put into, or out of mesh, signifying movement of them, axially or radially, into or out of engagement and the depth of mesh is determined by the centre distance between the two shafts.

Messenger

Endless rope passing from capstan to cable to haul it in. The small rope attached to a large wire or hawser to pull the latter between ship and shore.

Metacentric height

See *GM.* and *Figure 8.*

Methane (CH_4)

Residual LNG is mostly methane, a highly industrial and domestic fuel. See *LNG.*

Methyl chloride (CH_3Cl)

A colourless poisonous gas, chemical formula: CH_3Cl, also known as chloromethane. It is used as a refrigerant for cold storage.

Michell thrust block	Thrust block in which the thrust is taken primarily on segments that can cant slightly away from collars on the shaft.
Micro-seizure	The partial welding of materials in close contact caused by friction, as a result of which very small particles are torn away from the surfaces. See also 'fretting'.
Midship area	Immersed area of the midship section.
Midship section	Section at middle of ship's length.
Milling	The action of a milling machine on a work piece subjected to the cutting action of rotating cutters. The work piece may be fed horizontally or vertically into the cutters.
Millscale	Magnetic oxide of iron (Fe_3O_4) formed on steel at high temperature. It is formed on steel products in the processing mills and on solidification. It is very adherent to the basic metal.
Mimic diagram	Single-line system diagram usually attached to switchboards or control boards in which visual indicating devices show the operational condition of the various elements in the system.
Mineral oils	Named because of derivation from petroleum (petra – rock; oleum – oil) obtained from the earth. Almost universally used as lubricants as they are stable over wide range of temperatures and viscosities, readily available at moderate cost, giving protection against corrosion.
Misalignment	Departure from co-linearity of one or more parts of a machinery arrangement which should be in line.
Modulation	Method of using radio frequency carrier waves to transmit audio frequency signals. Achieved by varying either the frequency or the

amplitude of the carrier wave at audio frequency.

Module

(1) A basic length or ratio used as a means of comparison of different items, e.g.

$$\text{Module} = \frac{\text{circular pitch}}{\text{number of teeth}} \text{ for gear teeth.}$$

(2) A production limit, or component part, which is standardized to enable straightforward assembly, replacement, or exchange. Such units are much used in control and electronic equipment to enable repair by replacement following fault-finding routines to identify damaged modules. (3) In shipbuilding a module is a collection of machinery and equipment assembled for installation on board as a self-contained unit.

Molasses

Uncrystallized syrup drained from raw sugar; a cargo with problems for ship owners due to the high viscosity of the treacle.

Molecular

Applying to the molecule of a substance, that is the smallest amount which can exist in a free state.

Molybdenum (Mo)

Heavy metal element (sp. gr. 10.0) used as an alloying element to strengthen steel. It does not corrode in sea water. It is also used as a sprayed coating on steel to which it has good adhesion.

Moment to change trim (MCT)

A rough approximation is:
MCT by 1 cm = displacement/100 tm.

Monitoring (or surveillance)

The gathering of information concerned with the actual behaviour and performance of the plant. Facilities provided for this purpose may include equipment for indication, recording and alarm operation. Where centralized monitoring is provided, information is relayed to a central area such as a control room for ease of assessment by the duty engineer.

Mooring ring

Oval casting set in bulwark plating through which mooring lines are passed.

Mooring (single point)   The berthing or mooring of a ship's bow on to a floating buoy which is secured to the sea bed. The ship is then free to swing with wind or tide. The mooring buoy may have pipes which are connected to an underwater pipeline to enable the loading or discharge of an oil tanker without the need for port facilities.

Moulded breadth   Measured at midship section and is the maximum breadth over frames. See *Figure 2*.

Mud box   A strainer to intercept and retain insoluble matter and sludge.
     In the machinery space and shaft tunnel, the bilge pipe is led to a mud box which is accessible for regular cleaning.

Muff coupling   Used to connect two shafts without the use of flanged ends or coupled bolts. Consists of two sleeves tapered to fit each other. These sleeves fit over the two shafts and when the outer sleeve is forced into the taper, the inner sleeve is made to grip the shafts forming a solid coupling.

Muffler   American expression for exhaust silencer.

Multigrade oils   Lube oil containing additives known as viscosity index (VI) improvers fall into more than one SAE grade. They are designated by two extreme SAE Numbers, e.g. 10W/30, means a low temperature viscosity appropriate to the W grade at $0°F$ $(17.8°C)$ and a high temperature viscosity appropriate to the non W grade at $210°F$ $(98.9°C)$. Multigrade oils can therefore be used over a wider range of climatic air temperatures than ordinary lube oils. See *SAE*.

Muntz-metal   Old name for a copper alloy containing two parts of copper and one part of zinc by weight.

Muriatic acid (HCl)   Hydrochloric acid. A solution in water of the pungent gas hydrogen chloride, chemical formula: HCl. Also known as spirits of salts. It is used as a pickling agent to remove scale. Concentrated acid contains about 40% by weight of hydrochloric acid.

Mushroom valve	A tulip shaped valve made of heat resisting steel used for the inlet and exhaust valves on steam engines. It consists of a circular head with a conical face which seats over the inlet or exhaust port. A guiding stem also lifts the valve when moved by a rocker or tappet. Also known as a poppet valve.

N

Naphtha	Inflammable oil obtained by dry distillation of organic substance such as coal, shale or petroleum.
Navigator	Person who navigates or cons a vessel underway.
Needle rollers	Bearing rollers whose length is many times their diameter.
Negative earthed system	A d.c. system in which the negative pole is permanently connected to earth.
Neoprene	Polychloroprene rubber having fair to good resistance to petroleum-based fluids together with good resistance to ozone and weathering. Widely used for protective bellows and gaiters, etc.
Nest of tubes	Arrangement of a number of tubes, generally parallel to each other, used in heat exchangers to provide large area for heat transfer.
Net tonnage	This is given by a formula which is a function of the moulded volume of all cargo spaces of the ship.
Network analysis	Graphical method of planning a project in a logical sequence by plotting major activities with start/finish dates and times for each activity enabling overall time required to complete project to be estimated.
Network scheduling	The scheduling or time-tabling of production, assembly, despatch or some particular activity

by use of networks and network analysis techniques.

Neutral conductor   Any conductor connected to the neutral.

Neutral (system)   In 3-phase, star connected systems, it denotes the common point to which the corresponding end of each phase winding is connected, the other end of each phase winding being connected to a separate line terminal.

Nickel (Ni)   Ductile metal element (sp. gr. 8.9) used in pure form for electro-plated finishes and added to alloys to impart special properties. In steel increases susceptibility to heat treatment. In copper a range of alloys (cupro-nickel) are produced for their resistance to corrosion in sea water.

Nitriding   A means of surface hardening special steels. The steel is heated in an atmosphere of ammonia gas to a temperature of approximately 500°C. The depth of hardness depends on how long the steel is left at this temperature.

Nitrided Steel   Steel which has been subjected to an atmosphere of cracked ammonia in a sealed box at a temperature of 500°C for a specified time. The result is a very hard skin or case formed on the surface.

Nitrile rubber (or bunaN)   General term for copolymers of butadiene and acrylonitrile. Resistance to petroleum-based fluids varies from good to excellent according to the acrylonitrile content of the polymer. Heat resistance is moderate to good but resistance to ozone and weathering is generally poor. Nitrile rubbers are widely used for sealing duties in mineral oils.

Nitrogen (N_2)   Gaseous element which forms about four-fifths of the earth's atmosphere. It is an inert gas but can cause brittleness in steel if introduced during manufacture.

Node   (1) Point of rest in vibrating body. (2) Point

at which curve crosses itself. (*Figures 13 and 14*).

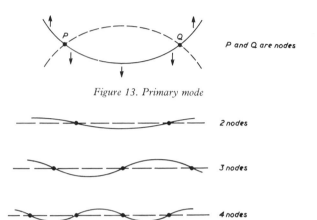

P and Q are nodes

Figure 13. Primary mode

2 nodes

3 nodes

4 nodes

Figure 14. Modes of vertical vibration

Nodule

Small round lump of anything, such as carbon in cast iron.

Noise (sound)

Noise is an unpleasant and unwanted sound. Sound is defined in BS661: 1969 as a 'mechanical disturbance, propagated in an elastic medium of such a character as to be capable of exciting the sensation of hearing'. This capability is determined by the frequency of the disturbance. The audible frequency range for a young person is approximately 20 to 20,000 Hz. The lower limit is difficult to determine because low frequency vibrations of air can be perceived although not heard in the normal sense of the word. The upper limit reduces with age (presbyacusis) and hearing damage. Frequencies below and above the audible range are known as infrasonic and ultrasonic respectively.

Non-destructive testing (NDT)

Processes for testing for properties, quality or soundness of materials or components, which can be applied without causing damage which would render the subject of the tests unserviceable. Typical non-destructive test processes include radiography, dye penetrant, magnetic particle, supersonic methods for detecting internal flaws or cracks, hardness testing, inspection by electron microscope and spot chemical tests.

Non-return valve

A valve in a pipeline which will automatically close if the direction of fluid flow is reversed.

Normalizing

Form of heat treatment for steel by raising its temperature above the upper critical point and cooling in still air. It results in grain structure refinement and improved mechanical properties.

Notch ductility

The ability of a material to withstand stress when the component which is made from it contains a notch or has a shape causing stress concentrations.

Notch effects

The reduction in the ability of a component to withstand shock loads due to the intensifying of the stress in the vicinity of the notch.

Notch steel

Steel having good 'notch ductility' properties has the effect of making it difficult for a crack to propagate. Notch ductility is a measure of the toughness of the steel.

Notch tough steel

Steel which has been heat-treated and/or alloyed to increase its ability to withstand brittle fracture (*q.v.*).

Nozzle

Outlet orifice for a pressurized fluid so shaped usually to convert the pressure energy into velocity energy and to cause the resulting jet to conform to requirements as to shape, concentration or dispersion, etc. Typical examples are the nozzles of injectors, ejectors, fire hoses, or nozzle forming passages which direct the flow of live steam or gases into the moving blading of a turbine. The term may also be applied when a ship's propeller is shrouded by a ring or duct so as to increase thrust at low speed; or by means of a pivoted nozzle ring, to direct the slip stream so as to improve rudder action when manoeuvring, particularly at low speeds or in confined waters.

Nozzle group

A segment of the complete nozzle plate in a

steam turbine containing a number of nozzles in a single nozzle plate. The nozzle group has its own separate section of nozzle box and its own nozzle control valve. For efficient operation the minimum number of nozzles should be open for a given power.

Nozzle propeller

See *Nozzle*.

N.P.L.

Abbreviation for National Physical Laboratory, a British Government laboratory for research in physics, metallurgy and kindred subjects. Situated at Teddington, Middlesex.

Nuclear Propulsion

A nuclear power reactor is simply a source of heat, on a par with a furnace of a boiler, with the important difference that several years' supply of fuel is contained in the reactor's core. The problem of nuclear power is to exchange the heat from the core and use it to make steam. Once the steam is generated it is treated exactly as if it were steam produced from an oil-fired boiler. The source of heat, the reactor core, is made up of sealed tubes containing uranium fuel pellets. The pellets are heated by the fission process which is controlled. The application of nuclear power to merchant ship propulsion is a developing subject which has not yet achieved any sort of finality but is widely used in submarines as it eliminates the problems of exhaust gas disposal.

Nucleate boiling

Occurs when the bulk of the fluid is at saturation temperature or slightly above. Large bubbles are formed on the hot surface which travel through the liquid to the free surface. The agitation of the fluid by the bubbles greatly improves the heat transfer rate.

Nucleus

A vital central point. A particle of matter acting as focal point from which reactions or changes of state may evolve.

Nylon

Polyamide thermoplastic materials characterized by their high strength and resistance to abrasion. They do not have rubber-like properties and in hydraulics are used mainly as bearings and bushes rather than for sealing.

O

'O' sealing ring	Toroidal ring made of an elastomer such as rubber used for sealing circular flange joints or the gap between piston and cylinder in pneumatic or hydraulic mechanisms.
OBO ship (Ore/bulk/oil)	This ship type has been designed to carry its full deadweight with dry cargo in bulk, such as ore, coal, grain, etc. or a liquid cargo such as crude oil.
Observation tank	A tank which receives drains from fuel oil heating, tank heating, and other possible contaminated steam heating lines. Also known as an observation drain tank.
Octane	Relates to fuel used in IC engines. The octane number or rating indicates the anti-knock properties of the fuel and is equal to 100 for one type of octane. The development of high octane has permitted a corresponding increase in engine compression ratios with good anti-knock properties.
Ogee ring	The steel ring which connects the bottom of the furnace to the shell of a vertical boiler.
Ohm	Unit of resistance in which a current of 1 amp flowing for 1 s generates 1 J of thermal energy. This unit is termed the Ohm (Ω).
Ohmmeter	Instrument measuring electrical resistance.
Ohm's law	The fundamental law in electric circuit theory is Ohm's Law which states that the current through any circuit element is proportional to the voltage across it. It can conveniently be expressed by $I = V/R$ where I = amperes; V = volts; R = ohms.
Oil, black	Applies to non-distillate crude oil, black oil products (including fuel oil, heavy diesel oil, bitumen and asphalt solutions) and lubricating oil.
Oil, diesel (heavy)	Marine diesel oil, other than those distillates of

109

which more than 50 per cent by volume distills at a temperature not exceeding 340°C when tested by ASTM Specification D86/67.

Oil fuel register

The arrangement of vanes, air swirler plates, etc. fitted between the inner and outer casing of a boiler at each burner position. The register regulates the supply of air to the burner to ensure complete combusion of the oil. Usually referred to as the air register.

Oil, non-persistant

An oil which evaporates in air leaving little or no trace, such as petroleum spirit and kerosene.

Oil, persistant

An oil which does not evaporate in air, such as crude oil and lubricating oil.

Oil separator

A device used to remove water and other impurities including some solids from lubricating or fuel oils. Normally of the centrifugal type although static types are available for dealing with oily ballast water.

Oil, white

Distillate oils which disperse easily and are less of a pollution hazard. Hydrocarbon oils other than 'black oil'.

Oily water separator

Term used to denote equipment designed to separate oil from oil-water mixtures in such a way as to achieve acceptably clean water.

Opposed piston engine

Internal combustion engine having two pistons in each cylinder. Charging, compression and ignition take place between pistons, each of which operates its own connecting rod.

Ore carrier

Singledeck ship designed to carry a homogenous cargo throughout the length. Machinery located right aft. Some ore carriers transport oil as an alternative cargo. See *OBO*

Orlop

Lowest deck in the ship.

Orsat apparatus

A portable apparatus for analysing flue or exhaust gases. A measured amount of gas is passed through three tubes successively, which contain potassium hydroxide to absorb carbon dioxide, pyro-gallol to absorb oxygen and the third, acid cuprous chloride to absorb carbon monoxide. The reduction in the volume of gas after it has passed through each tube indicates the amount of each constituent gas thus absorbed.

Oscillation

(1) The oscillation of a ship which if excessive can increase stresses in the structure and cause distress to passengers and crew. The principal ship oscillations are rolling, pitching and heaving. (2) A periodic change in a variable. An oscillation may have a decreasing, constant or increasing amplitude.

Osmosis

The passing of solvent through a semi-permeable membrane separating two solutions of different solute concentration. The phenomenon can be observed by immersing in water a tube partially filled with an aqueous sugar solution and closed at the end with parchment. An increase in the level of the liquid in the solution results from a flow of water through the parchment into the solution.

Otto cycle engine

The working cycle of a 4-stroke engine – suction, compression, explosion and exhaust. Two revolutions of the crankshaft per cycle.

Outboard

Suspended or projecting outside the ship.

Outrigger

Extension to increase spread of stays to topmast.

Overfall (Propeller)

Turbulence caused by the flow of a strong tidal stream over an abrupt change in depth or as a result of the meeting of tidal streams.

Overlap

Crank angle or period during which inlet and exhaust valves in an IC engine are open together. See *Lap* and *Lead*.

Override

Means for countermanding the effect of an automatic device such as an engine governor.

111

Owner

Registered owner or disponent owner, including a charterer by demise of a ship. A 'disponent owner' is the term for person or company to whom the ownership of the ship has been made over to, such as a company managing the ship for the owners.

Oxalic Acid $(2H_2O)$

Highly poisonous and sour acid found in wood-sorrel.

Oxidation

The reaction of a substance with oxygen. The chemical term is also used for a reaction which removes hydrogen from a compound or in which an atom loses electrons.

Most metals react with their environment and the result of this reaction is the creation of a corrosion product. Protection against atmospheric corrosion is important in ship construction not only at the building berth but also in the fabrication shops. Serious rusting may occur where the relative humidity is above 70%. All steel material in ship construction, plates and sections are shot blasted to remove rust and millscale. Shot blasting in shipyards takes the form in which the abrasive is thrown at high velocity against the steel surface and may be re-circulated. Following shot blasting the plates and sections pass through an airless spray painting plant.

Oxides

Chemical compounds resulting from the combination of substances with oxygen.

Oxter plate

Steel plate that fits around the upper part of the rudder post.

Oxygen (O_2)

Gaseous element which forms about one fifth of the earth's atmosphere. It is a reactive gas which supports combustion.

Ozone (O_3)

Condensed form of oxygen with molecular structure O_3 and refreshing odour. A powerful oxidizing agent formed when air is subjected to silent electrical discharge or to ultra-violet rays. It may cause damage to some cargoes, e.g. fruit.

P

Packing

Material closing a joint or assisting in lubrication of a journal. For normal water duties soft cotton packing is preferable. Packing may be impregnated with graphite or soaked in oil for initial running in. For chemical and oil duties asbestos and metallic packings are used.

Paint primers

Paints used on bare metal as a basis for subsequent coats of paint. The adhesion of the primer to the metal is particularly important. (1) Prefabrication primers are for the first coat on steel plate before it is fabricated into a ship's structure and should protect the steel from corrosion during the building period. (2) Wash primers contain a proportion of phosphoric acid which converts traces of rust to iron phosphate and forms an adherent base for subsequent coats. (3) Zinc based primers apply a thin coat of zinc which protects steel from corrosion.

Paints

Paint consists of pigments dispensed in a liquid referred to as the 'vehicle'. When spread out thinly the vehicle changes to an adherent dry film. The primary object of painting steelwork is to provide a coating that will protect the surface from the oxidizing effect of the air or water.

Panama chocks

Steel casting with oval opening. Chocks must be fitted at each end of ship passing through the Panama Canal, for use in the locks.

Panting

The pulsating in and out of the bow shell plates as the ship rises and plunges in the sea.

Panting beams

Additional beams in the forward and after portions of the hull to prevent panting action of the shell plating.

Panting Stringers

See *Panting beams*.

Pantograph

Instrument for copying a diagram on any desired scale.

Parabola	Plane curve formed by intersection of cone with plane parallel to its side.
Parallel body	See *Run* and *Figure 15*
Parallel connection	Connection in parallel is the converse of series, in that all positive terminals are joined together and all negatives are similarly joined. Current then passes from one joined set of terminals to the other, each component taking only a proportion of the total current. Term can be applied to lube oil and cooling systems.
Parallel operation	The connecting together of two or more power sources, e.g. alternators or boilers, such that the sum of their outputs is led to a common load.
Parbuckle	To roll a rounded object, e.g. a cask or a spar, up a ramp or up the ship's side by passing the bight of a rope under it, making fast the bight and hauling on the standing ends. May also be applied to careening or righting a vessel or boat, or to any operation involving rolling the object.
Parbuckling	The method by which a damaged ship lying on its side is righted. Two methods are used: one by the application of external forces which pull the ship upright and the other by an internal selection of combinations of flooding and buoyancy depending on the conditions. As well as the problem of rotating the vessel care must be taken to prevent the vessel going right over in the opposite direction.
Passive stabilizer	See *Stabilizer*.
Pawl	Pivoted catch engaging the teeth of a ratchet to permit rotation in one direction only as on a winch.
Pearlite	A micro-constituent of steel having a lamellar structure consisting of iron and iron carbide. It is formed when steel is slowly cooled.
Pedestal	Vertical member supporting a machine

component such as a bearing from a foundation below.

Peen

Wedge-shaped or thin end of a hammer head used to shape or mark metals by so doing to alter clearance.

Performance Monitoring (P.M.)

The automatic calculation and recording at predetermined intervals of critical plant performance criteria, such as specific fuel consumption, to monitor gradual changes in efficiency occurring in service. A valuable aid to watchkeepers for obtaining the optimum performance from an engine or machine. Up to 150 sensors can be fitted to a large diesel engine to record cylinder and fuel injection pressures, piston ring clearances, differential pressures and temperatures of key units etc. See *Condition Monitoring.*

Periodical survey

To maintain the assigned class the vessel has to be examined by the Classification Society's surveyors at regular periods.

Permeability

The percentage volume of a space that can be flooded is known as the permeability. When a compartment contains cargo, fuel, etc. the amount of water which can enter on damage is less than the volume of the empty compartment. The ratio of the volume entering to the volume of the empty compartment is called the permeability. For cargo spaces it is taken as 60% passengers and crew spaces as 95%.

Perpendiculars

Forward perpendicular (FP) is a vertical line through the intersection of the load waterline and the stem contour. After perpendicular (AP) is at where the aft side of stern post meets the load waterline or if no post at the centre of rudder stock. Length between perpendiculars (LBP) is horizontal distance between the AP and FP. See *Figure 5.*

Petroleum

Found in porous rocks and is almost always accompanied by gas. The principal producing land areas are North America, Venezuela, Arabian Gulf area, U.S.S.R., Africa and

Indonesia. In the last two decades the search for petroleum has been extended to offshore continental shelves and production has been developed in the Gulf of Mexico, Arabian Gulf and the North Sea.

Petroleum coke

A solid fuel with a low ash content which is non-clinkering and usually low in sulphur content. It is produced as a by-product in the cracking and distillation of petroleum.

pH value

The pH value of a solution is a convenient method of specifying its effective acidity or alkalinity.

Phase angle (θ)

Angle by which current lags voltage in an a.c. circuit containing resistance and reactance; or leads the voltage in an a.c. circuit containing resistance and capacitance. At zero phase angle, current and voltage are said to be in phase.

Phosphor-bronze

Alloys of copper and tin with a small addition of phosphorus which increases the strength of the bronze. A number of useful alloys are produced, with the amount of tin ranging from $3\% - 12\%$ which are resistant to corrosion in sea water.

Pickling

Process for removing oxide films from metals by dipping into a bath of acid. Sulphuric or hydrochloric are the most usual acids employed.

Pig's ear

A funnel or tun-dish. A pipe system may include an open-ended pipe which discharges the fluid being carried into a 'pig's ear' to give a visual indication of the flow of the liquid in the pipe system.

Pile

An assembly of moderator material such as pure graphite in which there is a neutron source, such as pure uranium, together with neutron detectors and counting equipment. These assemblies were originally built up by piling layers one on another and thus they were termed piles.

Pilgrim nut	Patent design of nut for securing propellers. The design incorporates a torus ring of elastomer embedded in the face of the nut which can be expanded under hydraulic pressure to force the propeller boss on to the tapered shaft end.
Pilgrim wire	Wire fixed at one end and passing over a pulley at the other with a weight hanging on the free end to maintain tension, used for checking alignment of main bearings. The catenary sag of the wire is calculated and allowed for when measuring position of each bearing.
Pillars	Supports to the deck.
Pilot valve	Small valve used to admit fluid to one side of a piston operating a large valve.
Pinion	Small cog-wheel engaging with larger one.
Pintles	Pins that hinge the rudder to the gudgeons on the rudder post.
Pipe-laying vessels	Ships designed to lay pipes on the seabed.
Piston end clearance	Minimum distance between the piston and cylinder closure in a reciprocating engine or pump.
Piston skirt	Cylindrical part of piston below the pressure rings keeping the piston in alignment with the cylinder.
Pitch	(1) The distance any specified point on the face of a propeller blade would move forward in one revolution. (2) The pitch of a screw thread is the amount of axial advance per revolution. Thus when a nut is turned round on the bolt for one revolution, the nut travels a distance equal to the pitch of the screw thread on the bolt.
Pitching	The action of a ship in moving to the crest and

descending into the trough of a wave. Movement about transverse axis.

Pitting

Corrosive action on steel plates making small surface indentations.

Planimeter

Machine for measuring the area within a closed plane figure.

Plasma coating

Deposition of substances on to components by means of a plasma gun or torch. Temperatures of about 30,000°C may be reached in the gaseous stream emitted from the torch so that most materials which are introduced into the stream in powder form are volatilized. Coatings of a variety of substances, metals and ceramics, can be built-up by the process.

Pleuger system

As conventional rudders are of limited use at low speeds, a way of providing manoeuvring capability at low speed is to deflect the propeller race. This can be done by using deflector plates or by turning the propeller disc itself. The latter is the principle of the Pleuger active rudder which is a streamlined body mounted on a rudder, the body containing an electric motor driving a small propeller. To derive full advantage the rudder should be capable of angles greater than 35°. With the ship's engines at stop, the system can turn the ship in its own length.

Plimsoll Line (or mark)

The Load Line for British cargo ships laid down in the Merchant Shipping Act of 1875. An international load line was adopted by 54 nations in 1930 and amended in 1968 to include a new line permitting a smaller freeboard in new large ships. Named after Samuel Plimsoll, a merchant and shipping reformer, who campaigned against the large number of ships that were lost each year at sea.

The calculated freeboard results in a load line which is boldly marked on the sides of the ship. Different marks are required as for Winter North Atlantic (WNA), Winter (W), Summer (S), Tropical (T), Fresh water (F) and Tropical fresh water (TF). See *Load line* and *Figure 12*.

Plummer Block	Apparatus for supporting revolving tail or propeller shaft with removable cover giving access to bearings.
Plunger	A part sliding in another, often cylindrical and sufficiently close fitting to retain fluid pressure as in 'fuel pump plunger'.
Pneumatic	Acting by means of compressed air. There are many types of both percussion and rotary tools operated by compressed air. Machinery control systems, etc.
Pneumercator gauge	Instrument for indicating depth of a liquid in a tank.
Polarization	Effect in a primary cell of the liberation of hydrogen at the surface of the copper electrode, which leads to loss of efficiency. It is avoided by the addition to the electrolyte of a depolarizing agent such as copper sulphate.
Pole (battery)	Terminal of positive or negative polarity, i.e. the positive and negative poles of a d.c. system.
Pole (rotating machine)	That part of a magnetic circuit which carries or in which is embedded an excitation winding or a permanent magnet.
Pollution (marine)	The prevention and control of marine pollution from ships has been a concern to IMCO since its inception. The first major step towards international control of marine pollution was taken by a Convention in 1954. The latest Convention covers pollution from ships by oil, noxious liquid substances, exhaust gases, noise, sewage and garbage. Oil-carrying ships must be capable of operating with a method of retention on board in association with the 'load on top' system. IMCO issues the publication 'Manual on Marine Pollution'.
Port	(1) Left side of a ship when looking forward. (2) Harbour designed to look after ships.

E

(3) Opening in the ship's side for goods or personnel.

Poop

The after superstructure on the upper deck.

Potentiometer

Instrument for measuring or adjusting electrical potential.

Preferential trip

Automatic switch fitted in an electric circuit to protect the main switchboard from overload. When the load current reaches a specified amount the trip operates cutting off the supply of electricity to non-essential auxiliaries only.

Pre-ignition

Uncontrolled burning of the combustible charge in an engine from a hot surface such as the exhaust valve, etc.

Pressure ring

Ring on a piston designed to prevent pressurized fluid leaking from one cylinder end to the other.

Preventers

Additional stays to support the mast.

Priming

(1) Boiler. Projection of minute particles of water into the steam due to impurities in feed water and incorrect water level in boiler. (2) Pump. Insertion of water to expel air and break air lock. (3) Painting. The first coat applied to a base surface assisting the adhesion of the final coating.

Prismatic self-supporting tanks

Special tanks adopted in LNG tankers. Such tanks of self-supporting type may be prismatic (single or double-walled), spherical or cylindrical.

Probe

(1) An instrument inserted into the interior of machinery or tubes for examination purposes. Electronic devices are available for drawing through a tube to indicate defects in or the thickness of the tube wall. (2) An exploratory bore made in a metal.

Products carrier

A tanker designed to carry refined products such as gas oil, aviation fuel, kerosene, etc. Large numbers of tanks, several separate loading and discharge piping systems and suitably coated tank surfaces are particular features of this type of ship. The purpose being the segregated loading storage and discharge of a number of separate parcels or types of products in any one voyage.

Professional engineer

A professional engineer is competent by virtue of his fundamental education and training to apply the scientific method and outlook to the analysis and solution of engineering problems. He is able to assume personal responsibility for the development and application of engineering science and knowledge, notably in research, designing, construction, manufacturing, superintending, managing and in the education of the engineer. His work is predominantly intellectual and varied, and not of a routine mental or physical character. It requires the exercise of original thought and judgement and the ability to supervise the technical and administrative work of others.

His education will have been such as to make him capable of closely and continuously following progress in his branch of engineering science by consulting newly published work on a world-wide basis, assimilating such information and applying it independently. He is thus placed in a position to make contributions to the development of engineering science or its applications.

His education and training will have been such that he will have acquired a broad and general appreciation of the engineering sciences as well as a thorough insight into the special features of his own Branch. In due time he will be able to give authoritative technical advice, and to assume responsibility for the direction of important tasks in his Branch. (Definition adopted by Engineering Societies of Western Europe and U.S.A. (EUSEC).)

Profile

Drawing showing elevation of ship indicating the location of decks, bulkheads, etc. See also *Lines plan.*

Propane (C_3H_8)

One of the petroleum gases obtained as a by-product in various oil refining processes. Colourless easily liquified gas found in natural gas. Important raw material for the petro-chemical industry. Boiling point –43.8°C. Readily transported as liquid in tanks and bottles. Used as fuel for domestic and industrial uses.

Propeller

The force needed to propel a ship is obtained from a reaction against the water causing a stream of water to move in the opposite direction. Of the devices used – oars, paddle wheels, jets and the screw propeller – the latter has almost exclusive application with seagoing ships. See also *Propeller efficiency*, *Propeller slip*, *Propeller law* and *Pitch (propeller)*. Fundamentally the marine screw propeller can be regarded as a helicoidal surface which on rotation screws its way through the water. A screw propeller has two or more fixed blades projecting from a boss. The surface of each blade when viewed from aft is called the face; it is the driving surface when producing an ahead thrust. Types of propeller: (1) Fixed pitch propeller. The blades in the form of a screw have a fixed pitch. (2) Controllable pitch propeller. The blades can be controlled hydraulically to vary the pitch and it is thus possible to provide astern thrust without reversing the direction of rotation. (3) Ducted propeller. The propeller is surrounded by a duct which can be moved into the vertical plane to provide helm movement. (4) Bow thrust propeller. A propeller mounted in a tunnel athwart the hull at the forward end of the ship to provide athwartships movement for manoeuvring. (5) Voith Schneider propeller. A patent design of propeller which rotates in a vertical plane. The vertical blades oscillate to provide thrust in the horizontal plane. (6) Keyless. Propellers are traditionally attached to the propeller shaft on a tapered cone with a key running longitudinally. Recently a device has been invented whereby a bush inside the propeller boss is expanded on to the taper by hydraulic pressure. When sufficient pressure is obtained, the hydraulic fluid is sealed off and maintains a frictional grip of the bush on the shaft without the need for a key. A successful joint requires exact machining to match both bore and cone mating surfaces. (7) Contra-rotating. Two propellers mounted on concentric shafts rotating in opposite directions. The arrangement balances the torque reaction generated by a single propeller. (8) Cupped. Cupping consists of a slight turn in the trailing edges of the propeller blades, to produce greater thrust and to reduce the onset of cavitation. They can be run higher on the transom, which reduces the drag effect of the lower unit. Cupped propellers are usually only fitted to light, fast boats; their pitch should be about one inch less than normal propellers.

Propeller, cavitation

See *Cavitation*.

Propeller efficiency

The ratio Thrust power/delivered power, i.e. PT/Pp is known as the propeller efficiency.

Propeller law

As with resistance experiments on model hulls, Froude's Law of Comparison may be used to predict the performace of geometrically similar propellers. The results are usually plotted in the form of non-dimensional coefficients.

Propeller, ringing

Before the onset of cavitation, the blades of a propeller may give out a high pitched note or ring. Singing is due to the elastic vibration of the material excited by the resonant shedding of non-cavitating eddies from the trailing edge of blades. Singing is a nuisance to the ship's crew and dangerous in warships as the high pitch can be picked up by enemy sonar. The problem is overcome in naval propellers by sharpening the blade edge as silent blades are essential at low speeds.

Propeller slip

The difference between the distance a screw would advance in one revolution in a solid medium and the distance it actually advances in a given medium.

$$\text{Slip in metres} = P - \frac{30.864\ V}{N}$$

where P = pitch in metres; V = speed in knots; N = rev/min.

123

Psychrometer Wet and dry bulb hygrometer.

PTFE (Teflon or Fluon) Polytetrafluoroethylene (PTFE) is a thermo-plastic polymer which is virtually immune to chemical attack and which may be used over a very wide temperature range.

Punkah louvre Ventilation system whereby fresh air can be directed at a given temperature to different parts of the ship. Air is delivered from a spout working on a universal joint which can be rotated to suit the individual's requirements.

Purging system System designed to free an enclosed space from inflammable mixture by flushing it with an inert gas or liquid.

Purifier Rotary machine used for centrifuging contaminants from fuel or lubricating oil.

Push rod Rod working in compression to operate one machine part from another. Used for example between cam follower and valve rocker in poppet valve engines.

Pyrometer Instrument for measuring very high temperatures.

Pyrotechnic Rockets and flares used for illumination, distress signals and target marking.

Q

Quadrant Fitting attached to the rudder head and connected to the steering gear.

Quadruple expansion Steam engine arrangement in which steam passes through four cylinders in series.

Qualification A document attesting that a person has fulfilled certain conditions either academic or practical or both and is therefore capable of holding a certain position.

Quality assurance A system of inspection and testing materials and components to ensure that they conform with the specified requirements. Planning such a scheme is particularly important when large numbers of similar articles are being produced and it would be impracticable to test each one individually.

Quality control

Inspection and testing of materials and components to ensure their uniformity and fitness for service.

Quarter deck (raised)

In the smaller deadweight ships the upper deck at the aft end is frequently raised to increase capacity. The height raised is less than that of the full height of a 'tween deck. In warships, the officers' deck.

Quarters

Living accommodation spaces. See also *Accommodation*.

Quenching

Rapid cooling of a metal. In the case of heat treatable metals, quenching is carried out to change the structure prior to further heat treatment to enhance their properties.

Quill shaft

Shaft connecting two rotating components of a machine designed to provide torsional flexibility or to permit misalignment or axial movement.

QPC

Quasi-propulsive coefficient. Given by

$$\frac{\text{effective power} + \text{allowance for weather}}{\text{delivered power}}.$$

R

Rack

Straight bar with teeth suitable for engaging with a pinion to convert linear to rotary motion or vice-versa.

Radar

A navigational aid which uses radio waves of very short wavelength sent out as a narrow beam by a highly directional aerial. The aerial rotates sending the beam out through a full 360°. Any solid object of reasonable size will reflect the beam and be detected on the radar screen as a bright spot.

Radial flow	Fluid flow outwards towards the periphery of a circular body or from the periphery towards the centre.
Radiographic flaw detection	Detection of flaws in materials by subjecting them to rays of very short wave-length which penetrate the material and are exposed on a photographic film placed on the opposite side. Flaws detected on the film by the differences in density of the film in way of the flaws. X-rays emitted from an electrical apparatus may be used or gamma rays from radio-active materials such as radium or isotopes of iridium, caesium or cobalt which themselves have been subjected to radio-activity.
Raised forecastle	Superstructure at extreme forward end of a ship. In certain circumstances may not be full 'tween-deck height.
Raised quarter deck	See *Quarter deck*.
Rake	The inclination from the vertical of masts, funnel, etc.
Ramp (Ro/Ro)	The success of the Ro/Ro ship depends greatly on the equipment to get vehicles from shore into the ship. The link for this is a ramp hinged at the ship end and supported at the outer end such that the ramp can be adjusted to suit tide levels.
Rankin cycle	Thermal cycle used as a standard for heat engines and heat pumps employing a condensable vapour as working fluid.
Ratchet	Toothed wheel or rack capable of movement in one direction only. Movement in the other direction being prevented by a pawl.
Rated output	The BSI issue two specifications about rating requirements. These are BS 649 – oil engine types and BS 765 for carburetted engines. Oil engines: the rated load is the load in brake power which the engine can carry for a period of 12 hours at rated speed under the following conditions:

Mean barometric pressure 75 cm of mercury
Atmospheric temperature 85°F (29.4°C)
Humidity 1.52 cm of mercury vapour pressure.

Ratline

Small lines fastened across ship's shrouds as ladder-rungs.

RAZ

Naval term to describe the operation of replenishment at sea when a ship is supplied with fuel, stores and ammunition whilst underway, from another ship or helicopter.

Receiver

Person or firm to whom cargo is consigned.

Reciprocate

Move towards and backwards cyclically over a fixed path.

Reciprocating pumps

Pumps of the positive displacement type. A piston working in a cylinder displaces a given volume of fluid for each stroke. The amount delivered depends on the piston area and piston speed. They are used for small quantity high pressure duties, have an efficiency of about 85 per cent and are self-priming.

Rectifier

Static device for converting electrical power in the form of a.c. to electrical power in the form of d.c.

Reducing valve

Valve arrangement designed to maintain constant fluid pressure at its downstream side irrespective of inlet pressure.

Reduction gear

Machine with two or more gear wheels of different sizes meshing together to reduce the output shaft rotational speed to less than that of the input shaft.

Redwood Seconds

Kinematic viscosity is expressed in Redwood seconds. Three types of apparatus are used: the Redwood in the U.K., the Saybolt in the U.S.A., and the Engler on the Continent. In each case the time in seconds for a given quantity of liquid to run from the viscometer is measured. S.I. unit is the centistokes $(cSt) = 10^{-6} m^2/s$.

Reef

(1) Ridge of rocks near surface of the sea.

(2) Line sewn into a sail and used to shorten sail by lacing to a boom.

Reefer

(1) Refrigerated ship specially engaged in carrying perishable food products. (2) Mid-shipmans jacket.

Reeve

(1) Pass a rope through a block. (2) Fasten rope block, etc.

Refractory

A material which is able to withstand high temperature conditions such as boiler brickwork.

Refrigeration

Is obtained by the vapourization of liquid refrigerants. Installation consists of evaporator, condenser and compressor. The refrigerants in marine installations are Freon 12 and Freon 22.

Register

Book in which particular information is entered; specifically applied to a ship's 'Certificate of Registry'.

Reheat

Used with reference to high pressure main boilers. The superheated steam is returned to the boiler after doing work in the first part of the high pressure turbine. In the boiler the steam temperature is increased, sometimes to the initial temperature, at constant pressure. The steam is then led back to the next stage of the turbine plant and improves the efficiency of the plant.

Reliability

Is based on the premise that equipments fail in service if run for long enough. Even the most reliable equipment will fail in time not necessarily due to wearout. If a number of identical equipments are monitored over a long period a pattern of failures against time will occur. These failures will show up as a 'bath tub' curve with three distinct regions of failure; a number of failures at the start, reducing in number as the machine settles down to a steady failure rate rising again at the end of its useful life.

System and equipment failures must be defined before a reliability analysis can be carried out to establish the failure rate and the mean time between failures (MBTF).

Relief valve

Valve loaded by spring or other means to release fluid from a system when the pressure in the system reaches a preset level.

Residue

The waste or low value substance remaining following a process, e.g. the oil–water emulsion remaining after tank cleaning.

Resilient mountings

Machine support fixing embodying springs or elastomeric material permitting limited movement of the machine relative to its foundation to prevent the transmission of vibration.

Resistance

(1) The total resistance of a ship moving in a calm water surface has a number of components, wavemaking, frictional, form drag, eddy-making, air resistance and appendage resistance. Skin friction is the resistance due to roughness of the hull or surface of a body in a gas or liquid. (2) Extent to which the flow of current in a body is restricted. It is represented by the quotient of a given direct voltage at the terminals of the body and the current passing through it. Ohm's Law: $V/I = R$, where V = voltage, I = current and R = resistance.

Retraction

1. A withdrawing or removing action. 2. (U.S.) Astern movement from a beach.

Reverse current protection

Type of protection for direct current only in which a release permits a mechanical switching device to open, with or without delay, when the current flows in reverse direction and exceeds a pre-determined value.

Reynold's number

The point at which laminar flow changes to turbulent flow is governed by a constant called Reynold's Number which is given by $\frac{VL}{\gamma}$ where V = speed relative to still water; L = length and γ = coefficient of kinematic viscosity of the fluid.

Rhumb

Angular distance between two successive points of compass.

129

Rhumb-line	Curve on the surface of map or globe cutting all meridians at the same angle. Navigationally applied to the track of a ship sailing in a fixed direction.
Riding lights	Common term to denote statutory anchor lights which a ship must carry when riding at anchor.
Rigger	Person who attends the wires and other rigging for the working of a ahip.
Rise of floor	See *Dead rise* and *Figure 2*.
Rising main	A pipe running vertically from a fire main to carry a water supply either from pumps or to hydrants.
ROB	Round of beam. See *Camber* and *Figure 2*.
Rockershaft	A shaft which carries rockers to transmit the activating force from the camshaft to open the valves in an internal combustion engine.
Rockwell hardness test	An indentation hardness test using a diamond cone (Brale) or steel ball. A load is applied to the indenter to force it into the surface of the article to be tested. The depth to which the indenter travels into the surface is read on a clock gauge mounted on the machine and is a measure of the hardness of the material.
Rolling	Rotational motions about a longitudinal axis are called rolling. See *Swaying*.
Rolling keel	See *Bilge keel*.
Ro/Ro	Roll-on/roll-off ships are in general designed for the carriage of motor cars, commercial motor vehicles and unitized cargo.
Rope guard	Shaped steel plate attached to the stern tube boss of a ship which extends from the boss towards the propeller leaving a small enough gap between the two to prevent ropes from becoming wrapped round the shaft. They are usually made in two semi-circular halves and bolted or welded to the stern tube boss.

Rotary pumps	Vane pumps which differ from centrifugal pumps in that they are true displacement pumps and differ from reciprocating pumps in having a higher rate of leakage. Suitable for small quantity medium pressure pneumatic systems which provide lubrication, where viscosity helps to reduce leakage in hydraulic systems and for large deliveries at low pressure. Similar in principle to a gear pump in which two specially shaped members rotate in contact.
Rotating machine	Electric apparatus depending on electro-magnetic induction for its operation and having components capable of relative rotary movement used for converting electrical energy into mechanical power, e.g. an electric motor.
Rotor	Revolving part of a rotary machine such as a turbine or alternator.
RPM	Revolutions per minute (rev/min), or rotational speed of machinery and shafts.
Rubbing strake	Longitudinal strake or stringer fixed outside the ship's skin plating and positioned so as to make first contact with harbour walls, or other vessels in event of impact or rubbing between them. Any resulting damage to the rubbing strake is repairable without affecting the ship's structure or plating.
Rudder	All ships must have some means of directional control and generally this is achieved by means of a rudder. The three main types adopted are: (1) Hinged, unbalanced; (2) Semi-balanced; (3) Balanced. See *Figure 1*.
Rudder bearing	The mass of the rudder may be carried by a lower pintle and partly by a rudder bearing within the hull. In some designs the total mass is borne by the bearing.
Rudder stock	Vertical rudder shaft which connects to the steering gear.
Rudder trunk	The rudder stock is carried in the rudder trunk which has a watertight gland at the top where the stock enters the intact hull.

Run

The immersed body aft of the parallel body. The point where it joins the parallel body is referred to as the after shoulder. The parallel body is that portion of the underwater body which lies amidships and for which the cross-section is constant. (*Figure 15*).

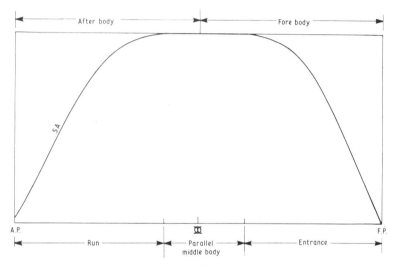

Figure 15. Parallel middle body

Runner

Rope in single block with one end round tackle block and other having hook.

Runout

Amount by which a rotating part is out of true or alignment.

Rust (Fe_2O_3)

Normal corrosion product of iron or steel in an atmospheric or water environment. It is the red oxide of iron (Fe_2O_3) referred to as white rust.

S

Saddle tanks

Tanks, usually for water ballast, which 'saddle' or are fitted over the upper sides of the main cargo tanks. They are triangular in longitudinal cross section and increase in depth towards the ship's side. They are fitted principally on bulk carriers and provide a means of increasing the ship's centre of gravity when a light cargo is being carried.

SAE

System devised by Society of Automotive Engineers (SAE) in U.S.A. Used almost universally for classification of crankcase and transmission oil by viscosity, e.g. crankcase oils classified in 3 grades (SAE 5W, 10W, 20W) according to viscosity at 0°F (17.8°C) and 4 grades (SAE 20, 30, 40, 50) at 210°F (98.9°C). Transmission oils have 2 grades (SAE 75 and 80) at 0°F (17.8°C) and 3 grades (SAE 90, 140, 250) at 210°F (98.9°C). There is a viscous range common to both crankcase and transmission viscosities and an oil within this range can have 2 SAE numbers.

Safe working load (SWL)

The maximum load in working that should not be exceeded. Ship derricks under Factory Act regulations are initially tested to a proof load in excess of the specified SWL.

Safety valve

A valve which opens automatically in the event of excess pressure in a container, e.g. a boiler steam drum. When fitted to high pressure systems, the valve is designed to open quickly and snap shut to avoid wire drawing and damage to the walve seat. A safety valve is normally fitted with easing gear. There are statutory as well as insurance companies' requirements which must be strictly observed. These demand that every boiler has two safety valves and that they be mounted directly on the shell or the steam drum.

Sagging

When a vessel drops at the middle of the length. The opposite to 'hogging'.

Salinity

The degree of sodium chloride content in water.

Salinometer

Instrument for indicating the proportion of salt in a given quantity of water.

Samson post

See *King post*.

Sand-blasting	Blast cleaning with dry sand. Wet sand-blasting uses a sand and water mixture and is not so severe as dry sand-blasting. Great care must be taken to prevent sand entering and damaging machinery.
Sanitary pump	A sanitary system supplies sea water for flushing water closets, urinals, etc. A continuously operating pump is fitted to serve the system.
Saturation	(1) Condition of an electrical element where any further increase in its input signal produces negligible change in its output signal. (2) Steam at the same temperature as the water from which it was formed as distinct from steam which was subsequently heated becoming superheated steam. (3) Term can be applied to electronics, solutions, steam, vapour, magnetic materials and electrical currents.
Save all	(1) A device to prevent waste or spills under a machine or filter. (2) Strip of canvas laced to foot of sail. (3) Sheet of canvas to serve same purpose as cargo net.
Scale	Hard deposit that forms on inside of boilers or on exposed ferrous metals.
Scantlings	Thickness of plating, dimensions of ship frames, beams, girders, etc.
Scavenge	The process in an engine of evacuating the cylinder of combustion products and replacing them by fresh air.
Schottel rudder	Retractable rudder propeller powered by diesel engine.
Scoop condensers	Condensers which receive their cooling water from a scoop system. See *Scoop cooling*.
Scoop cooling	A scoop which protrudes below the level of a ship's hull plating and faces forward. The forward motion of the ship forces water into the scoop and through the cooling water system. The arrangement reduces the power

required for the main sea water circulating pump while the ship is moving ahead.

Scoring

Formation of grooves in a rubbing surface by foreign bodies carried by the mating component or by local seizure. See *Gear tooth damage*.

Scotch boiler

A series of furnaces connected with a common combustion chamber at the back of the boiler. To this chamber are linked a series of tubes in which the gases generated in the furnace pass.

Scraper ring

A form of piston ring fitted to the crankcase end of the piston designed to remove excess lubricating oil from the cylinder walls.

Screw

See *Propeller*.

Screw lift valve

A valve in which the plug is connected to the spindle and remains in the position set by the movement of the spindle.

Scrubber

Cleaning tower in an inert gas system. Sea water is sprayed into the gas path to remove soot and other contaminants before the inert gas is led to the cargo tanks or holds.

Scuffing

(1) Surface deterioration of steel due to lack of lubrication causing partial welding of two mating surfaces, e.g. gears. (2) Caused in diesel engines by metallic contact between piston ring and liner. When the load exceeds the mild/severe transition load plastic deformation occurs with the production of a plate-like metallic debris which is the main contributor to very high wear rates. See *Gear tooth damage*.

Scupper

Drain from deck to carry off rain or sea water.

Scuttle

(1) As a noun, an opening in a deck or ship's side, or a small hatchway affording access to spaces below, or a cover such an opening. (2) As a verb, to sink a ship by opening the sea cocks or blowing holes in the ship's bottom plating.

SD14

The Liberty replacement ships appeared in the

late 1960s and the two most popular were the
Japanese-built Freedom and the British SD14
designed and built by Austin and Pickersgill.
Over the years the latter has been very
successful and indeed the SD14 has become a
brand name in the world of shipping. Number
built – 180 to 1976; 15,000 dwt tonnes; general
cargo.

SDNR valve

Screw down, non-return valve.

Sea anchor

A device for keeping a small boat's head into the
wind and sea. One type used consists of a cone
shaped canvas tube open at both ends. A hawser
or warp is fastened to ropes from the sea anchor
which enable it to operate opened out by the sea
entering the cone. A tripping line is fastened to
the other end for hauling in. The sea anchor is
also known as a drogue or drag anchor.

Sea cocks

Valves which control the ingress of sea water
from inlets in the ship's bottom or at any point
below the waterline.

Sea inlet

Aperture in the ship's plating below the
waterline to permit sea water to enter a
pumping system for washing down decks,
removal of boiler ash, firefighting etc.

Sea tube

That part of the inlet end of a ship's sea water
circulating system which is attached to the
hull.

Seakeeping

The qualities that embrace the aspects of ship
design that affect the ship's ability to remain at
sea in all conditions and carry out its specified
duty. These include strength, stability and
endurance.

Seam

Lengthwise edge joint of any plating.

Seatings

Structural supports for main propelling
machinery, auxiliary machinery in engine
room and on deck.

Seaworthy

A ship is said to be seaworthy when it is fit in all
respects to carry the cargo and crew in good
condition, as far as protection from the action

of the sea, wind and weather is concerned, and deliver cargo at port of destination.

Segregated ballast
Sea water used as ballast which is not in contact with any cargo tanks. It is loaded into ballast-only tanks and discharged by its own separate piping and pumping system.

Self-trimming
Vessel with large hatches and clear holes that permit coal, grain and similar cargoes to be trimmed into any part of the hold.

Semiconductor
An electronic conductor, usually solid, whose resistivity lies between that of metals and that of insulators and in which the concentration of electrical charge carriers increases with increasing temperature over a certain range. For ordinary temperature ranges semi-conductors in contrast to metals have a negative temperature coefficient of resistance. Commonly used semiconductors include germanium, silicon and selenium.

Send
(1) Impulse given by wave. (2) Plunge or pitch of vessel due to impulse of wave.

Sensors
(1) A method of securing stability control on hydrofoil vessels. Can be untrasonic, electro-magnetic or feeler arms. A patent method is the Hydrofin. (2) Units for indicating pressures, temperatures and velocities.

Separator – oily water
Machinery used to remove oil from water which is to be pumped overboard. It consists of a tank, which may have filters and baffles, in which separation of the different liquids takes place, mainly due to the effect of gravity. The oil is led to the dirty oil or sludge tank and the water pumped overboard.

Series connection
A number of electrical components, such as batteries, resistances and capacitors, are said to be connected in series when each positive terminal is joined to the corresponding negative of the next component so that the current passes through them in succession, each component taking the full current. Term also applies to fuel oil and cooling systems.

Series wound

Denotes that a rotating electrical machine has excitation supplied from windings which are connected in the primary series circuit, carrying either the whole or part of the load current.

Serve

Bind with small cord to prevent fraying.

Servo-motor

Final controlling element in a servo-mechanism. It is the motor which receives the output from the amplifier elements and drives the load. The motion produced can be rotary or rectilinear as in the case of a hydraulic piston and cylinder. Hydraulic electrical or pneumatic power can be used to operate the servo motor.

Setscrew

Screw threaded for the full length of the shank for locking a collar, sleeve or coupling onto a shaft.

Settling tank

Oil fuel must be provided to the service tank in an uncontaminated state. As oil and water separate naturally the foregoing is achieved by providing deep settling tanks which permit this to occur in time.

Sewage treatment

The reduction of raw effluent into an inert sludge and water suitable for discharge overboard. The treatment may be chemical or natural.

Sextant

Reflecting instrument used for measuring altitudes and other angles.

Shackles

'U' shaped steel forgings with a pin through an eye on each end of the 'U' which serve as connecting links for rigging components. Used to anchor one end of a leaf spring to take up length variation on deflection.

Shaft alignment

Main propulsion shafting is checked for alignment during installation by use of a taut wire, theodolite or laser measuring techniques. When in service the shafting can be checked for static and dynamic alignment. The former can be done with clock or strain guages and the latter by measuring the amplitude of the axial vibrations when the ship is underway at various speeds.

Shaft generator

Generator not having an individual prime mover, but which is driven from the main engine or screw shaft by belt, gearing or other means. Generators driven from a ship's propeller shaft have economic advantages but may have to be disconnected when the ship is manoeuvring or when steaming at reduced powers.

Shaft Horse Power

(s.h.p.) is the net power delivered to the shafting of an engine after passing through gear boxes, thrust blocks, etc. It is measured on the shaft usually by a torsion meter.

Shaft tunnel

When the propelling machinery is not located fully aft it is necessary to enclose propeller shafting in a watertight tunnel between aft end of machinery and aft peak bulkhead. It protects shafting in way of after cargo holds.

Shale

A kind of naptha.

Shale oil

Oil distilled from oil-bearing shale-type geological formations of a clay rock which can be split readily into thin laminae.

Shank

(1) The straight part of a bolt or, rivet inserted in the hole of plate to be connected. (2) Part of the drill which is held by the drilling machine.

Shear force

At any section of a material is the algebraic sum of all forces on either side of the section.

Shear stress

A material is subjected to a shear stress across a surface within it when forces tend to make the part on one side of surface slide past the other part.

Sheave

The running wheel of a block.

Sheave block

Blocks are classified as single, double, triple, etc., according to the number of sheaves and are divided into two main groups, those for wire rope and those for manilla rope.

Sheer	The longitudinal curvature of the deck at side between the ends of the ship. See *Figure 5*. It is the height of deck at side above a line drawn parallel to the base and tangent to the deck line at amidships. The so called 'standard' sheer is given by Sheer at AP (cm) = 0.833 L(m) + 25.4 FP (cm) = 1.666 L(m) + 50.8 The sheer forward is twice the sheer aft.
Sheerstrake	The strake of shell plating at deck level.
Sheet	(1) A piece of metal, paper or other material of reasonable area but relatively thin section or thickness. (2) Any line or wire rope which is fastened to the lower after corner of a sail or boom to enable setting or trimming of the sail. (3) A sheet anchor is a third anchor of heavy construction carried by sailing ships for use in heavy weather.
Shell bearing	Thin walled steel shell lined with anti-friction metal. Usually semicircular and used in pairs for main and big-end bearings.
Shell expansion	Plan showing the disposition and thickness of all plates comprising the shell plating.
Shell plating	The plates forming the watertight skin of the ship.
Shelter deck	Superstructure deck continuous from stem to stern.
Shelter decker	Due to a legal discussion in the nineteenth century the space between the uppermost continuous deck – shelter deck – and the second deck was under certain conditions regarded as an open space. This exempted the 'tween-deck space from inclusion in the tonnage measurement. These ships had limited draughts and the hull scantlings were reduced accordingly.
Sherardizing	Patented process for coating small articles with zinc consists in placing the articles, after degreasing, etc. into a drum containing zinc dust. The drum is then rotated in a furnace at a

temperature of about 370°C and a matt grey coating is then produced on the surface.

Shifting boards

Portable bulkhead members fitted fore and aft in cargo holds when carrying grain etc. that might shift when vessel is rolling.

Shim

Thin strips or washers, usually of brass, copper or steel, used to take up or to adjust clearances either endwise on shafts or diametrically on split journal bearings.

Ships

The ship is one of the oldest methods of transport and apart from the ships built for special purposes, merchant ships are indeed part of a transport system. Their function is thus to transport commodities or people from one place to another. Ship types developed over the years have been dictated largely by the types of cargo. The General Cargo Carrier was a single or two deck ship with machinery amidship and with four or five holds. Today the tendency is to position the machinery further aft so that there are three or four holds forward and one hold aft of the machinery space.

In the transportation of general cargo there are four major modes of unitized carriage of goods. (1) roll on/roll off; (2) pallet ships; (3) barge carriers; (4) container ships. Of these the container ship has been given the most attention in terms of investment.

'Bulk carriers' are single-deckers with propelling machinery right aft. They can be loaded without having to trim the cargo.

'Oil tankers'–single deck ships with in general two longitudinal bulkheads. Deadweight up to about half a million tonnes.

'Liquefied gas carriers'–Liquid natural gas (LNG) consists largely of methane and for liquid form the temperature has to be reduced to 165°C. This means that special materials are required for the tanks containing the liquid. Liquefied petroleum gas (LPG) is carried in tanks in fully refrigerated ships.

'Container ships'–the cargo is stowed in 'boxes' or containers of standard size which are readily loaded and unloaded. Rapid turn-rounds are thus possible. The vessel is a single-decker with machines towards the after end.

'Refrigerated cargo ships' or 'Reefer' ships – As transporters of perishable goods the holds and 'tween-deck spaces are insulated and there is a refrigerating system that maintains these spaces at low temperatures.

'Passenger ships' – There is a wide variety of different types in this category which ranges from car and passenger ferries to cruise liners.

'Special ships' – Have been developed and can be classified as Special Transport – Icebreakers, hovercraft, hydrofoil vessels, multi-hull ships.

'Subservient transport' – Cable ships, dredgers, tugs, light vessels.

Sea-going fishing – Trawlers

Service vessels – Drilling ships, supply vessels.

Shock valve

A relief valve set to lift on the application of a sudden load. When fitted to hydraulic steering gears, the valve will lift if the rudder is hit by a heavy sea, allowing the rudder to move and preventing damage. The hunting gear will return the rudder to its original position.

Shot-blasting

Blast cleaning with shot. Steel or cast iron shot may be used, the latter being more abrasive. Care must be taken to prevent abrasive dust entering and damaging adjacent machinery.

Shrouded propeller

See *Nozzle Propeller*.

Shrouding

(1) Shrouds. A set of ropes from the mast of a ship to the gunwale to support the mast. (2) A strip of metal to which each turbine blade is attached, usually by riveting a tenon (*q.v.*) at the tip of the blade to the strip.

Shrouds

Wire rope which extends from mast head to vessel's side affording lateral support for the mast. Shrouds are often dispensed with and preventers adopted when heavy derricks are used. The mast has adequate scantlings to remain unstayed.

Shunt wound	Denotes that a rotating d.c. machine has excitation supplied from windings which are connected across the whole or part of the primary series circuit.
Shuttle valve	The valve used in the control of the supply of steam to the cylinder of a double acting single cylinder reciprocating pump. The shuttle valve moves from one end of the valve chest to the other, normally in the horizontal direction. This motion is controlled by the action of an auxiliary valve.
Side scuttles	(1) Side lights. (2) Portholes in the sides of ship and casings to give light and air.
Side keels	See *Keelson*.
Side rods	Tension rods used in opposed piston engines to carry load from the upper pistons to the side crossheads and thence through connecting rods to the crankshaft.
Side stringer	Fore and aft girder running along the inside of shell plating.
Silica gel	Amorphous form of hydrated silica. Used in the form of hard granules, it is chemically and physically almost inert but highly hygroscopic and thus very effective as an absorbent of fluids or vapours. Most usual uses are for drying air, dehydrating gases or filtration. After use, it can be re-generated by heating to drive off the absorbed matter. Used with inhibitor to dehumidify cargo holds.
Silicon (Si)	Brittle metal element (sp. gr. 2.4). Used as an alloying element in cast iron, steel and many non-ferrous metals and added as ferro-silicon in the manufacture of steel to assist de-oxidation.
Sill	(1) The height of an opening above the deck. (2) Upper edge at bottom of opening into a dock.
Simpson's rules	Rules to determine area of a plane figure bounded by a straight line, two perpendiculars and a parabolic curve.

Singing propeller	Resonant vibration of propeller blades due to the wake formation. The vibration may give rise to a high pitched whine which can be heard in the ship. Silent propellers essential in warships to avoid detection.
Single deck	Vessel having no deck below the weather deck.
Siphon	Pipe shaped as inverted 'U' with unequal legs for conveying liquid over edge of container and delivering it at a lower level by utilizing atmospheric pressure. Requires priming before flow can commence.
Sisal	(1) Hydroscopic material that can absorb much moisture and become very heavy; such materials may also swell and cause atructural damage. (2) Term used to describe sisal-hemp, or agave fibre, used for ropes, matting, fenders, etc. Somewhat cheaper and less satisfactory than true hemp because it is less flexible and swells more when wet.
Skeg	Finlike projection on bottom of vessel to support lower edge of rudder.
Skew gear	A bevel gear arrangement where the driven and driving shafts are at right angles but in different planes or skew. Also termed hypoid bevel gear or skew bevel gear.
Skin friction	See *Resistance*.
Slag	Metal arc welding started as bare wire welding; it was found that by dipping the wire into lime a more stable arc was obtained. Many forms of slag are now available for coating the wire or for deposition on the joint prior to welding.
Slamming	In severe pitching, the bow of the ship may leave the water and when it plunges back again 'slamming' occurs. A large instantaneous force is generated near the bow which can damage the hull unless ship's speed is quickly reduced.
Sliding foot	Supporting foot for a machine part which permits linear movement, generally to accommodate thermal expansion as in a steam turbine.

144

Slime	A living fungus which relies on coolness, darkness and moisture to exist wherever a supply of decaying organic matter is to be found. The main group are called myxomycetes.
Sling	Apparatus used to support hanging mass.
Slop tanks	A tank (usually a pair) fitted on oil tankers at the after end of the cargo tank section. The slop tank provides a collecting place for oil and water mixtures resulting from tank cleaning operations.
Slow steaming	The operation of a ship at a speed below the normal operating speed. It is usually done for economic reasons in times when there is an oil glut or shipping surplus.
Sluice valve	Shut-off valve in which closure is effected by a member sliding across a pipe normal to the pipe axis.
Smoke detectors	A fire-detection system that enables the presence of a fire in the inaccessible places of the vessel to be observed.
Snatch block	A single block used to alter the direction of pull on a rope. Block with hinged side to permit fall to be put over sheave without reeving the end through.
Snifting valve	Valve for releasing gas from a hydraulic system.
Socket head screw	Screw with hexagonal socket in head for an Allen key.
Soft-nosed stem	Radiused plate above the waterline to form the upper part of the stem. In the event of a collision the plate will buckle under load and reduce the impact damage.
SOLAS	Safety of Life at Sea. A Convention as the result of Conferences called by IMCO, an organization created by the United Nations.
Solenoid	Consists of a number of turns of wire wound in

145

the same direction so that when the coil is carrying an electric current all the turns are assisting one another in producing a magnetic field.

Solid injection

Process of injecting liquid into a chamber by hydraulic pressure without the assistance of air or steam.

Solid state

A solid-state device is one in which the electron flow takes place within solids. For example, a transistor, logic element or integrated circuit.

Sonar

Sound navigation ranging. See *ASDIC*.

Sounding pipe

Pipe leading near to the bottom of an oil or water tank and used to guide a sounding tape or jointed rod to measure the depth of liquid.

Spalling

See *Gear tooth damage*.

Span block

See *Blocks*.

SPC

Self-polishing copolymer. Patent anti-fouling paint.

Special area

Sea area where particular care is required for recognized technical reasons in relation to its oceanographical and ecological condition and to the particular nature of the shipping traffic therein, to minimize oil pollution, e.g. Mediterranean Sea, Baltic Sea, Black Sea, Red Sea and the Gulfs area, as defined in Regulations 10(1) of Annex 1 of the International Convention for the Prevention of Pollution from Ships, 1973.

Spectacle frame

Casting which projects from the ship's sides to take the bearings of the propeller shafts of a twin-screw ship.

Spiked cargo

One in which a certain amount of liquefied natural gas has been injected during loading. Should not be loaded into a normal tanker as due to boiling the gas will raise the pressure in the cargo tank.

Spindle

A pin on which anything rotates or a pin used

to operate some device e.g. a valve spindle when turned opens or closes a valve.

Spiral gears

When teeth cut on a helical plan are designed to connect skew shafts, they are known as spiral gears. Theoretically, they have point, instead of line, contact and there is a longitudinal sliding motion between the teeth.

Spline

Rectangular key fitted into grooves of a shaft.

Split bearings

Bearings made in two or more pieces to permit dismantling without axial withdrawal.

Spreader

(1) Iron bar or wooden spar fastened to a mast to increase the angle at the head of standing rigging and thus improve the staying. (2) Iron bar or framework used to separate the legs of wire slings when lifting a large load.

Springing

Essentially a continuous vibrating response sometimes sustained over long periods. It is a 2-node vertical vibration which can occur with short wave lengths and augment the wave induced stresses considerably.

Springs (moorings)

The lines which keep a ship from drifting fore and aft when moored.

Sprinkler system

To put a fire out it must be cooled sufficiently or deprived of oxygen. A method of cooling a fire is by an automatic sprinkler system actuated by bulbs which burst at about 80°C permitting water to pass.

Sprocket

Toothed wheel transmitting power by a link chain.

Spur gear

Term usually applied to gears in which the teeth are straight and parallel to the axis and similar in profile throughout their length. The gears are cylindrical in form and used to connect parallel shafts. Spur teeth may be cut externally on a cylindrical blank, as for an ordinary pinion; internally, as for an annulus; or on a straight rack to form a rack-and-pinion mechanism.

147

Spur wheel	Toothed wheel used for transmitting power between parallel rotating shafts. See also *Spur Gears.*
Spurling pipes	The port and starboard anchor cables are fed into the appropriate chain locker compartment through chain pipes or spurling pipes. These pipes are of tubular construction with castings as end mouldings to prevent chafing.
Squirrel cage motor	Type of induction motor in which the rotor consists of bars permanently short-circuited through stout end rings.
Stability	Tendency of a ship to remain upright or the ability to return to the upright when listed by the action of waves, wind, etc.
Stabilizer	Used to reduce the rolling motion of a ship. The systems fall under two main headings – passive and active. Passive: Bilge keels, fixed fins, tank systems. Active: Active fins, tank systems, moving weight.
Stable	The act of being stable, i.e. the stability, is the tendency of a ship to remain upright or return to the the upright position when inclined by a disturbing force. For stability to exist the centre of gravity (G) must be below the metacentre (M) See also *GM–metacentric height* and *Figure 8.*
Stack	See *Funnel.*
Stanchion	Vertical column supporting deck, guard rails, etc.
Starboard	The right hand side of ship when looking forward. Opposite to Port.
Stay	Wire rope from mast head to deck to prevent mast bending.
Steam	Water which has been converted into a vapour. Heat is applied to the water, at its saturation temperature, to change its state from liquid to vapour. The steam may be wet, dry saturated, superheated or de-superheated.

Steam atomization	The breaking down of fuel into very fine particles by the mixing of steam and oil in the burner nozzle to improve the combustion of the fuel and the efficiency of the plant.
Steam generator	(1) A piece of machinery used to produce steam from water. Used mainly in connection with lower pressure systems. (2) The package system of a steam turbine or steam engine connected to an electric generator.
Steaming slow	See *Slow steaming*.
Steering gear	The machinery used to turn the rudder on receiving a signal from the wheelhouse or from some other steering position.
Steering station	Place on the bridge which is outfitted with a steering mechanism and a compass and from which the helmsman steers the vessel.
Stern frame	Large casting, forging or fabricated structure at aft end of ship with, in general, vertical rudder post, propeller post and aperture for propeller.
Stern tube	Forms the after bearing for the propeller shaft and incorporates the watertight gland where the shaft passes through the intact hull.
Stiff	A vessel with a large metacentric height may have a short period of roll and thus roll uncomfortably. The opposite to 'tender'.
Stiffening	Steel sections used to stiffen bulkhead or other plating.
Stirling cycle	A closed circuit external combustion engine cycle in which air is compressed, heated, allowed to expand and then cooled before being compressed again. By using heat from the expanded air to preheat the external combustion air, high efficiency is obtained. Engines using this cycle are quiet and can burn low grade fuel but tend to be bulky. See also *Stirling engine*.
Stirling engine	An engine in which work is performed by the

expansion of a gas at high temperature to which heat is supplied through the cylinder wall. See also *Stirling cycle*.

Starter (fluorescent lamp) Automatic device which by interrupting the current to an inductive choke applies, for a short period, an initial voltage high enough to cause ionization of the gas within the lamp.

Starter (motor) Combination of all the switching means necessary to start and stop the motor, together with suitable overload protection.

Stator Stationary portion of a rotating electrical machine which includes the magnetic parts with their associated windings.

Stealer plate Single wide plate butt-connected to two narrow plates.

Steam drum That part of a water-tube boiler which receives the steam from the generator tubes.

Steel An alloy of iron and iron carbide. The carbon content is usually below 1% except for special steels. It is amenable to heat treatment and to alloying with many other metals to give alloys with a wide range of properties.

Stellite Trade name of the Deloro Stellite Co. Ltd. for a series of alloys used for services where hardness and wear resistance are required. They are basically alloys of cobalt and chromium with additions of tungsten and other elements.

Stem The structure at the extreme fore end of the ship rising upwards from fore end of keel. See *Figure 5*.

Stevedore Person or firm contracting to load or discharge cargo.

Stirrup Machine component shaped like a 'U' with a shaft attached across the open end co-operating with a bearing in another component.

Stirrup pump Developed to meet the threat of fires caused by incendiary bombs. It is a simple double acting pump in which there are only three moving parts: (1) the plunger rod, (2) the ball acting as a foot valve, (3) a ball forming a non-return valve in the piston at base of plunger rod.

Stock anchor

Has a straight shank with arms forged or cast as an integral part. Now replaced by stockless anchor.

Stockholm tar

A kind of tar prepared from resinous pinewood and used to preserve ropes from effects of water.

Stockless anchor

Consists of a head with flukes, a shank attached to the head and a shackle fastened to the other end on the shank.

Storm oil

Vegetable or animal oil used to hold down a breaking sea during an emergency situation such as hoisting or lowering lifeboats.

Strain, gauge

Instrument for measuring the amount of deformation of a material when subjected to stress. The most commonly used is the electric resistance strain gauge in which fine platinum wire is bonded to the component. When the component deforms due to applied stress, the wire is stretched which alters its resistance to the passage of electric current through it. The change of resistance is proportional to the amount of change of length of the wire.

Strake

A course of plating on shell, decks, bulkhead, etc.

Strength deck

Deck designed as the uppermost part of the main hull longitudinal strength girder.

Stringer

The strake of deck plating at the ship's side.

Stripping pump

Some tankers require a stripping connection from each tank to the cargo oil main. The stripping pump discharges to the riser.

Stroboscope

Instrument for observing moving bodies by making them visible intermittently and thereby giving them the optical illusion of being stationary.

F

Stroke	Distance travelled by the piston of a reciprocating engine or ram in moving from one end of the cylider to the other.
Strop	Length of rope with ends spliced to make a loop.
Strum box	Perforated plate box welded to the open end of bilge suctions in holds and other compartments. The box prevents debris being taken up by the bilge pump.
Stud welding	A shielded arc process. The stud is inserted in a gun chuck and a ceramic ferrule is slipped over it before the stud is placed against the plate surface. When the arcing period is complete the stud is driven into a molten pool of weld metal and thus attached to the plate.
Stuffing box	(1) Sleeve round a pipe or rod to prevent escape of steam or water. (2) Fitting at aft peak bulkhead in way of tail shaft.
Submersibles	Vehicles suitable for deep diving applications such as repair work on undersea cables. One such type is the *Pisces*.
Suction pressure decay	The fall off or reduction in the net positive suction head (n.p.s.h.) of a pump with increasing flow rate.
Sullage stripping	To remove residues, emulsified mixtures and sludge.
Summer tanks	Tanks on outboard sides of expansion trunks when oil tankers had such an arrangement. They were reserve spaces that could be filled with oil to bring vessel to the loading mark in summer.
Supercargo	Person in merchant ship management.
Supercharging	The purpose of supercharging is to raise the density of the charge in the cylinder at the end of the suction stroke so that a greater weight of mixture or air is trapped in the cylinder. Most diesel engines are supercharged by means of an exhaust-driven turbo-blower.

Superconducting machines

The electrical resistance of a metal decreases, as the temperature decreases, approaching zero at absolute zero temperature ($0°K$ or $-273°C$). A current induced by a magnetic field in a ring of metal at such a low temperature will continue to circulate through it after the magnetic field has been removed. The phenomenon may be used to drive an electric motor with the minimum of electrical energy supplied to it.

Superheated steam

If heat is supplied at a constant rate to ice at, say, $-20°C$ and no heat is lost by conduction, convection or radiation then the temperature first increases uniformly from $-20°C$ to $0°C$ as indicated by line AB in Figure 16. It remains constant at $0°C$ for time BC to melt into water. Further heat raises the temperature uniformly to $100°C$ as shown by CD. If the pressure on the water surface is atmospheric the water boils and the temperature remains constant at $100°C$ until all the water is evaporated. If the heat supply is maintained after all the water has been evaporated, the temperature and volume of the steam will increase above their saturation values and the steam will become saturated as shown by EF. Superheating also occurs if steam is compressed without any loss of heat or if steam is throttled to a lower pressure.

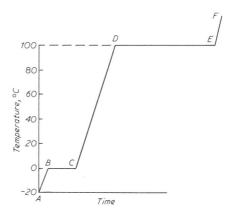

Figure 16. Steam temperatures

153

Superheater	A bank of tubes in the exhaust gas path from the boiler furnace. The steam from the boiler steam drum passes through the tubes where its temperature is raised, at constant pressure, above the saturation temperature.
Superstructure	Decked structure above the upper deck and the outboard sides of which are formed by the shell plating.
Surge	In still water at constant power a ship will move at constant speed. On encountering waves there will be a mean reduction in speed due to the increased resistance. The speed is no longer constant and the term surge defines the variation in speed about the new mean value. See *Surging*.
Surging	The fore and aft linear movement is called surging. See *Swaying*.
Surveyor	Official of a classification society or of Department of Trade who reports that a ship and the machinery conform to the standards of the rules of the society and recommends Certificate of Class to be issued.
Surveys	See *Periodical surveys*.
SVP	See *Vapour pressure*.
Swaging	A process of forging metals to reduce the cross-section of the piece. The workpiece is held between dies which are hammered, so that its sides remain concentric. The process is used for reducing the diameter or tapering of bar stock.
Swash plate	(1) Longitudinal or transverse non-watertight plate fitted in tank to reduce swashing action of liquid contents. Its function is greatest when tank is only partially full. (2) A disc set obliquely on a shaft so that it wobbles as the shaft rotates. Axially arranged pistons and cylinders may be operated by the swash action and this mechanism is commonly used in hydraulic and fuel pumps to control delivery.
Swaying	A ship moving in a wave system has six degrees

154

of freedom, three linear and three rotational: heaving, surging, swaying, rolling, pitching and yawing. The transverse movement – linear – is called swaying.

Swedged plating Corrugated plating that gives rigidity to the plating without the use of stiffeners.

Synchronizing Process whereby the voltage, frequency and phase angle of an incoming alternator are adjusted to be as close as possible to those of the machine or system with which it is to operate in parallel. The process is carried out after the incoming machine has been run up to speed but before it is connected to the system.

T

Tabernacle Socket for hinged mast to permit, say, passing under a bridge.

Tachometer (1) Instrument for measuring velocity (usually of rotation). (2) Name given to 'counter' indicating number of propeller revolutions per minute.

Tail shaft The part of the propeller shaft that passes through the stern tube and takes the propeller. Normally a short section of shafting to facilitate removal of shaft for repairs.

Tallymen or checkweighers Those who keep an account of intake and outturn of cargo.

Tank cleaning Cleaning out and clearing of gases from the cargo tanks on a tanker. Automatic mechanical tank washing machinery is used to jet high pressure hot or cold water around the tank. Portable or fixed machines may also be used which enter the tank through circular apertures in the deck.

Tank tests Tank tests of a model hull are generally specified. These provide the designer with a range of speeds and corresponding powers for the hull form. Due to the very considerable costs involved in conducting full scale experiments on the resistance of ships a great deal of attention has been given to experiments on models. The resistance of a ship and the power required to drive it can be estimated from the model.

Tank top

Plating forming the top of the double bottom.

Tappet

Sliding piston-like cam follower with either a flat end or a roller bearing on the cam.

TBN

Abbreviation for total base number. Applies to lubricating oils and indicates the amount of alkali (base) available to neutralize acids.

Tchebycheff's rules

Used to determine area of a figure bounded by a straight line and a curve. The ordinates do not require internal multiples as in Simpson's Rules.

TDC

Top dead centre. The point of highest travel of a piston in its cylinder.

Technician

The technician is competent by virtue of his/her education, training and practical experience to apply in a responsible manner proven techniques and procedures and to carry a measure of technical responsibility, under the guidance of a technician engineer or a chartered engineer. He/she requires the ability to communicate clearly, both orally and in writing. The technician should possess a recognized academic qualification, exemplified at present by an Ordinary National Certificate or City and Guilds Part II Certificate in appropriate subjects.

Technician Engineer

The technician engineer is competent by virtue of his/her education, training and subsequent experience to exercise independent technical judgement in and assume personal responsibility for duties in the engineering field. His/her education and training is such that by the application of general principles and established techniques, he/she is able to understand the reasons for and purpose of the operations for which he/she is responsible. The Technician engineer performs technical duties of an established or novel character either independently or under the general direction of a chartered engineer or scientist. He/she requires the power of logical thought and when in a management role the quality of leadership. His/her work is at a higher level than that of a technician. The technician engineer has an academic qualification of a standard not lower than that exemplified at present by a Higher National Certificate or a City and Guilds Full Technological Certificate and has had a minimum of five years' engineering experience of which two years must have been devoted to practical training.

Telemotor system

An arrangement of hydraulic pipes for controlling a ship's rudder. As the helmsman rotates the wheel a plunger is moved which applies a control pressure to the steering gear control valve via hydraulic pipes which in turn move the rudder. Some modern ships are fitted with an electrical instead of hydraulic control system.

Tell tale

(1) Index adjacent to steering wheel to indicate position of rudder. (2) Instrument in Master's cabin for checking ship's course. (3) Method of indicating engine orders to/from bridge.

Tempering

Heat treatment process usually carried out after quenching to impart ductility to metals which become brittle on quenching.

Template

Mould or pattern.

Tenon

A protrusion at the end of a component which fits into a socket in another component for attachment. Tenons are provided at the tops of turbine blades to attach them to the shrouding.

Tensile test

Test in which a specimen of known area of cross-section is subjected to increasing stress in tension until it fractures. Measurements taken are: stress to fracture (ultimate tensile stress (UTS)), yield point or elastic limit, elongation at fracture (measured between gauge marks), reduction of area (the difference between the cross-sectional area of the specimen before testing and the area at the fracture).

Terotechnology

The name given to the process of co-ordinating and controlling the various techniques and practices which influence the efficiency and profitability of capital plant, equipment and engineering installations. Embracing a number of separate but related techniques, sciences and disciplines, terotechnology is defined as 'the technology of installation, commissioning, maintenance, replacement and removal of plant, machinery and equipment, of feedback of information to design and operation thereon, and of related subjects and practices.'

Thermal loading

Stress level resulting from heat flow into the walls surrounding a hot zone in a machine such as a heat engine.

Thermal trap

Device operated by temperature variation for releasing water from a steam system.

Thermistor

Resistor having a very large non-linear and usually negative temperature coefficient of resistance. Frequently inserted between the windings of rotating machines to form an embedded temperature detector or safety indicator.

Thermo-couples

When certain metals are joined and the junction heated, an electric current is generated, the potential of which can be measured away from the heat source by a millivoltmeter (the cold junction). The potential is related to the difference in temperature between the hot and cold junctions.

Thermometer pocket

Recess in a machine or pipe to take a thermometer. To ensure accurate readings the recess should be kept full of oil which acts as a medium for heat conduction.

Thermostat

(1) Device for maintaining or adjusting

temperature. Expandible element may be used either to cut off heat supply or to start flow to cooling medium when temperature exceeds required value. (2) Chemically, a bath which is maintained at a constant temperature.

Thimble

Metal ring with concave sides into which a rope may be spliced to prevent rope fraying or weaving.

Thin shell bearing

Bearing in which bearing metal is bonded to a thin shell of stronger material which is enclosed in and supported by a heavier and stronger housing.

Three island ship

Ship having a poop, bridge and forecastle.

Throttle valve

(1) A valve used to reduce the pressure of a fluid, e.g. in a steam system it may be used to reduce the pressure of steam before it is used for auxiliaries. (2) A valve used to control the admission of fuel to an internal combustion engine.

Thrust bearing or washer

Bearing or washers for reducing friction through axial loading on rotating shafts or components.

Thrust block

Unit that takes the propeller thrust. As the thrust is of a pulsating nature the block foundation must be rigid.

Thrusters, bow and transverse

See *Propellers* and *Bow thrusters*.

Thyristor

A transistor in which one of the three electrodes (the control electrode) initiates the main current flow between the other two but does not limit it. The device is used as an electronic switch. The terms silicon controlled rectifier and thyristor are synonymous.

Time charter

The hiring of a vessel for a period of say months or for the time required for a specified voyage. In a gross time charter the owner pays for insurance of the ship, wages, stores, etc. The charterer must pay for bunker fuel, fresh water, stevedoring, port charges, pilotage, etc.

Timing	The point or time in a cycle when a particular operation takes place, such as a valve opening or closing, or fuel being ignited by a spark.
Titanium (Ti)	Ductile metal element (sp. gr. 4.5) which, when alloyed with small amounts of other elements, has a high strength to weight ratio. It is practically incorrodible in sea water. It is also added to some stainless steels.
Tonnage	See *Gross tonnage*.
Tonnage hatch	See *Tonnage length*.
Tonnage length	Term deleted under the International Conference on Tonnage Measurement of Ships 1969. The tonnage of a ship consists of gross tonnage and net tonnage. The unit is cubic metres and the word 'tons' no longer applies.
Tonne	Measure of mass in S.I. units (1,000 kg).
Topping lift	Wire rope extending from derrick head to mast for purpose of supporting the load.
Torque	The product of a turning force and the radius at which it acts is the torque about the axis of rotation, in Nm.
Torque convertor	Device in a drive line providing more or less torque in the driven shaft than in the driving shaft at correspondingly lower or higher rotational speed. The most common type is hydraulically operated.
Torsionmeter	Instrument that measures the power being transmitted by a shaft.
Torsion test	A test in which a specimen is fixed at one end and a torque load is applied axially at the other end. The load required to fracture the specimen and the angle of twist may be measured. It is most commonly used for testing bar or wire material for the subsequent manufacture of springs.
Torsional vibration	Cyclical twisting and untwisting of a shaft due

160

to variation in applied torque. When the frequency of torque variation coincides with the natural frequency of the shaft system resonance occurs and the vibration amplitude builds up to a higher than normal level. This condition is known as a critical speed.

Total energy

The total heat energy available from the combustion of the fuel used in an engine. A total energy system is one which attempts to use all the otherwise waste heat such as that in exhaust gas or cooling water.

TPC

Tonnes per centimetre. The mass to be added or deducted from a ship to change the mean draught by 1 cm. See also *Immersion*.

TPI

(1) Number of threads per inch. (2) Load in tonnes required to increase a ship's draught by one inch.

Transducer

(1) Electrical device for converting mechanical or thermal stress into an electrical signal for operating a warning lamp indicator, or measuring a load/stress. (2) The part of a sonar set or echo sounder which transmits or receives the sound signal into or from the sea.

Transformer

A static electromagnetic unit consisting of two windings magnetically linked by an iron core so that an alternating electromotive force applied to one of the windings will produce by induction a corresponding e.m.f. in the other winding. The windings are called primary and secondary. Transformers are used to convert energy supplied at one voltage into energy supplied at a different voltage.

Transistor

A semiconductor device capable of providing amplification and having three or more electrodes. A transistor consists essentially of a thin slice of P type or N type semiconductor between two slices of N type or P type semiconductor. The first combination is referred to as an N-P-N device and the second a P-N-P device. The terms P type and N type describe how the charge carriers flow in a semiconductor material; in P type material the hole density exceeds the conduction electron density, whereas in N type material the conduction electron density exceeds the hole density. A non-vacuum electronic device that can replace the thermionic valve.

Transmitter

A device designed for mounting at or near the point of measurement, which receives the primary measurement signal and generates a related output signal which is suitable for transmission. (1) Electrical component for transmitting the state of a measuring device, such as a fuel gauge, to a suitably scaled indicator. (2) Apparatus necessary for producing and modulating radio-frequency current. Includes associated antenna system. (3) Cylinder and piston unit in ship's steering gear telemotor system which transmits helmsman's wheel orders to rudder.

Transom stern

Stern of flat athwartship plates which offers increased deck area compared with the cruiser stern.

Transverse beam

See *Beams*.

Transverse framing

The side shell framing may be transverse or longitudinal, transverse being adopted in many conventional cargo ships.

Tribology

The study of what happens when surfaces 'rub' together, embracing friction and wear in all their aspects including lubrication, bearing design, and selection of materials. The term was introduced because other terms already in common use refer to only part of the technology of interacting surfaces. In industry it relates principally to bearings, slides, gears, brakes and clutches; in medicine and surgery mainly to diseased and artificial replacement of joints.

Trigger pulse generator

Circuit configuration designed to produce pulses of current such as would, for example change a thyristor from the off-state to the on-state.

Trim

The difference in draught forward and aft.

Trimming tank	Watertight compartment that can be filled with water ballast to change the trim of the ship.
Tripping bracket	Reinforcements on deck girders, etc. to prevent free flanges being deformed.
Trunk	Space formed by bulkheads or casings around openings through which access can be obtained without disturbing adjacent spaces.
Trunk Piston	Piston in a reciprocating engine connected directly to the crankshaft through a connecting rod without a piston rod or crosshead.
Truck to keel	Mast top to keel.
Tufnol	Synthetic plastic material used for bearings such as rudder pintles and stern tubes.
Tumble home	The inclination inboard of the upper sides of the ship. See *Figure 2*.
Tungsten (W)	Heavy metal element (sp. gr. 19.3). One of its chief uses is as an alloying addition to steel (12–18%) to produce high speed steel for cutting tools.
Tunnel	See *Shaft tunnel*.
Turbine	Basically a machine giving rotary motion obtained by steam/gas or water impinging on a vaned wheel.
Turbine Blading-end tight	A method of keeping steam leakage to a minimum between the pressure stages of a turbine. The shrouding is ground to a knife edge allowing a small axial clearance between the fixed and moving blades. This clearance can be varied, in some cases, while manoeuvring, to prevent damage to the blades. The clearance will be adjusted to a minimum after full away to obtain maximum efficiency from the turbine. Movement of the rotor is achieved by means of an adjustable thrust housing.
Turbine, gas	An engine in which the heat energy in a gas is

163

converted to rotational energy by the gas impinging on a series of blades on revolving discs. The gas turbine unit normally consists of compression, heating and expansion cycles.

Turbine, steam

An engine in which the heat energy in steam is converted to rotational energy by the steam impinging on a series of blades on revolving discs.

Turbo-blower

Multi-stage rotary air compressor. Often used incorrectly to describe a 'turbocharger'.

Turbocharger

A turbine driven air compressor powered by the exhaust gases from the parent internal combustion engine. The compressor has a compression ratio of up to 1:3:5, increases the power developed by the I.C. engine by up to two and a half times, rotates at 10,000 to 24,000 r.p.m. and is a major source of noise ($+100$ dB). In highly rated engines with high b.m.e.p's ratios of up to 1:6 may be required in which case two turbo chargers are fitted in series to provide two stage turbo charging but this arrangement is not widely fitted today.

Turbometer

Instrument for indicating the main engine revolution. Fitted in the engine room.

Turbulent flow

The movement of water inside the wake may follow two patterns; one is laminar and the other turbulent. In the latter eddying occurs. Reynolds investigated the flow of water in pipes, and demonstrated that there are these two distinct types of flow. With laminar flow in a straight tube the particles of fluid move in straight lines parallel to the axis of the tube. In turbulent flow the motions of the particles are in the form of spiral eddies.

Turnbuckle

Sleeve or link having right and left-handed screw threads at opposite ends fitted with corresponding eye bolts so that rotation of the body or sleeve either draws together, or separates, the eye bolts. A device much used for setting up and retaining tension in shrouds, stays or other rigging.

Turn-down ratio	Ratio of maximum to minimum flow through a device such as a fuel burner.
Turning circle	Standard manoeuvre carried out as a measure of the efficiency of the rudder.
Turning gear	Mechanism for turning the shaft of a machine for overhaul or setting purposes.
Turn of the bilge	Curved section between the bottom and side of the ship. See *Figure 2*.
Twaddell or Twaddle	Scale for the measurement of specific gravity of acids, etc. Used for checking the brine density in refrigeration systems, distillers and boilers.
Twaddel hydrometer	Used in testing the strength of tanning extract and the density of brine in refrigeration systems, distillers and boilers.
Tween-decker	Ship with more than one deck and thus provides space between two adjacent decks.

U

ULCC	The large tankers are crude oil carriers and known as Ultra-large Crude Carriers.
Ullage	(1) The quantity of a tank or oil compartment lacks being full. (2) Depth of space above the free surface of the fluid.
Ullage plug	At this point the ullage volume is assessed.
Ultrasonic testing	One of several methods of non-destructive testing. Used in locating defects in welds, particularly fine cracks.
Ultra-violet crack detection	Cracks in metals may be revealed by coating the metal with a penetrating oil which carries a fluorescent substance. The fluorescent oil

penetrates cracks which become bright under ultra-violet illumination.

Under deck tonnage See *Tonnage length.*

Underway Generally means that the ship is making way through the water. Technically is applied to a ship that is not attached.

Union purchase The commonest derrick rig is the 'single swinging derrick'. Adjacent derricks may be used in 'union purchase', the derricks being fixed in the overboard and inboard positions. Cargo is lifted from the hatch and swung outboard by controlling the winches for the cargo runner. See *Figure 3.*

Universal joint Enables the transmission of power by a shaft at any selected angle. See *Hooke's joint.*

Unmanned machinery space (UMS) Attained by remote control of ship machinery and its automatic operation.

Upper deck Uppermost continuous deck.

Uptake Metal casing connecting boiler or engine with funnel.

V

Vacuum Space from which air has been almost exhausted by air pump etc.

Vacuum circuit breaker Circuit breaker whose main current carrying contacts are enclosed within a highly evacuated envelope.

Vacuum contactor Contactor whose main current carrying contacts are enclosed in a highly evacuated envelope.

Valence Replacing power of an atom as compared with standard hydrogen atom.

Valency Unit of combining capacity. Chemical

equivalent mass = relative atomic mass/valency.

Value analysis

The systematic examination of products, component by component, with the object of minimizing cost without impairing function.

Value engineering

An approach to component design with the object of getting value for money. The cheapest materials and simplest arrangement are sought to obtain effective production. Simple assembly and minimum costs must also be aimed for.

Valve bridge

Part of mechanical valve through which valve spindle fits. As spindle is rotated, valve is raised or lowered via female thread in bridge.

Valve, by-pass

Device which permits unfiltered fluid to by-pass the filter element when a preset differential pressure is reached.

Valve chest

Suction connections are led to each hold or compartment from the main line. Valves are introduced to prevent one compartment being in direct communication with another. These screw-down, non-return valves are provided in a valve distribution chest.

Valve, electronic

Vacuum or gas filled tube whose electrical characteristics are governed by electron conduction between electrodes.

Vanadium (V)

Metal element (sp. gr. 6.0) which is added in small quantities to many high strength steels. It is present in some boiler fuel oils as vanadium pentoxide (V_2O_5) which produces a corrosive slag.

Vapour

Gaseous form of a normally liquid or solid substance.

Vapour pressure

The pressure exerted by the vapour from liquid in an enclosed space above the liquid. The vapour pressure, or more precisely the saturation vapour pressure (S.V.P.) varies with temperature.

Variometer	Device for varying the inductance in an electric circuit composed of two or more coils; the relative position of the coils being varied to each other.
Veer and haul	To pay or ease out cable and then immediately to haul it in. This allows machine to pick up speed.
Ventilator cowl	Hood-shaped top to a ventilator.
Venturi	Restriction or 'choke' in a tube, or pipe to produce a change in velocity and pressure. Technique used in carburettors and ejectors to suck out liquids, and in instruments to measure fluid flows.
Vernier	A pair of adjacent scales for determining very small measurements or making adjustments as in a micrometer or timing shafts.
Vessel bridge	Superstructure on upper deck from which a clear view forward and on either side is obtained and from which the ship is conned and navigated. In the three island ship the bridge is admidship and extends out to the ship's side, the side of the erection being a continuation of the side shell of the ship. In bulk carriers and oil tankers the bridge is aft. In some special ship types, drilling and supply vessels the bridge and control station is forward. See also *Bridge*.
Vessel bridge aft	When forward bulkhead of the wheelhouse is located at a distance from the stern which is less than 35% of the vessel's length.
Vessel bridge amidship	When the bridge is located amidship.
Vessel bridge forward	When the forward bulkhead of the wheelhouse is located at a distance from the stem which is less than 25% of the vessel's length.
Vessels, LASH	Special ships for the transport of barges. The scheme enables a number of barges which ply on inland waterways to be docked in the specially constructed mother ship, for transport across the open sea to another port,

168

where the barges are discharged. They can then continue on another waterway system, if necessary, to their destination without intermediate unloading and reloading.

Vibration (types)

Mainhull vibration is concerned with the entire structure vibrating and there are two types namely flexural vibration where the hull bends like a beam and torsional vibration, the structure twisting about a longitudinal axis. Flexural vibration is the most important. Vibration of hull structure has three principal sources: (1) the propellers, (2) the main propelling engines, (3) local due to auxiliaries. All parts of the hull girder have their natural periods of vibration and when that of machinery synchronizes with the ship girder then vibration may become serious. See also *Vibration-whipping*.

Vibration (mode of)

When a body in stable equilibrium is displaced by an external force it will ultimately return to that position when the force is removed and will oscillate. Consider a bar of constant section in the type of vibration in which the centre of gravity does not move. The bar then vibrates as a free beam. The ends move as the middle moves down as shown in Figure 13. This is called the primary mode as it has the least number of nodes – a node is a point of rest. A body can vibrate in different ways involving 2, 3, 4 or more nodes as shown in Figure 14. The natural frequency of the vibration increases as the number of nodes increase. The time of one vibration from one extreme position to the other extreme position and back again is called the period. The frequency is the number of vibrations in unit time and is the universe of the period.

Vibration-whipping

A ship struck by a slamming force will vibrate in its natural frequency and thus will eventually come to rest. High stresses can result from this whipping action and the stresses are frequently referred to as 'whipping stresses'.

169

Vickers hardness test	An indentation hardness test using a diamond square pyramid indenter which is forced into the prepared surface of an article under a standard load. The load is applied automatically by the machine and the diagonal width of the impression is measured by a microscope attached to the machine. The hardness number is calculated by dividing the load by the area of the impression or, in practice, by reference to tables.
Victory ship	Prefabricated U.S. built cargo ship of 1939–45 war period. See also *Liberty ships*.
Virtual inertia factor	In ship vibration the total mass being vibrated equals the displacement of the ship plus a mass of entrained water. The ship displacement may thus be considered to be multiplied by a factor called the virtual inertia factor to give the total virtual mass.
Viscometers	Three types are in use: Redwood, Saybolt and Engler. In each case the time in seconds for a given quantity of liquid to run from the viscometer is measured.
Viscosity	The internal resistance of a fluid to relative movement. The viscosity of fluids decreases as the temperature rises and is expressed in Redwood seconds. (1) Dynamic Viscosity of Poise (P): defined as tangential force exerted per unit area on each of two parallel plates at unit distance apart in the oil, one moving in its own plane with a unit velocity relative to the other. (2) Kinematic Viscosity of Stoke (St): for most purposes more convenient to use and is dynamic viscosity density. Units normally used are centipoise (cP) and centistokes (cSt), being a one-hundredth part of the respective unit. The most important property of a lubricating oil.
VLCC	Very Large Crude Carrier of 210,000 tonnes or over. Leap in size from 45,000 to 70,000 tonne tankers into the VLCC class started in the early 1960s when the European oil consumption increased eight-fold between 1950 and 1968. The benefit from large tankers was proved

after the blocking of the Suez Canal when the World Scale Index rocketed from 30 to 225 immediately after the war settling at 100 for the rest of the year. Ships had to travel round the Cape of Good Hope and the economics of size were proved. Much of the new technology was developed by Shell International Marine and involved the introduction of high tensile steel plates and improved corrosion control systems to save weight, large cargo tanks, development of single-screw 27,500 H.P. turbines with large single boilers, and a high degree of automation which reduced the size of crews. See *ULCC*.

Voith Schneider   Form of vertical axis propeller where the propeller blades are vertical and of aerofoil shape.

Volute   A term most usually applied to springs in which successive coils are of progressively increasing (or decreasing) diameter and arranged also with some spacing axially. The resulting geometrical arrangement may be described as a conical spiral, or conical helix. It is a spring configuration which results in a spring having a progressively increasing rate with increasing deflection and also facilitates having a large number of coils within a moderate overall length, successive coils being able to enter each other as compression proceeds.

Vulcan clutch or coupling   A hydraulic unit for connecting an engine to the propeller shaft consisting of an outward flow water turbine driving an inward flow turbine fitted in a common casing; similar to a fluid clutch in a motor car.

Vulcanite   Hard vulcanized rubber.

Waist   Upper deck between forecastle and poop. Term little used now.

Wake   A ship in motion carries with it a certain mass

171

of water. This 'wake' as it is called has a forward velocity in which the propeller operates and thus the speed of the propeller through the wake water is less than the speed of the ship.

Wall-sided ship

Vessel having vertical sides in the vicinity of the waterline.

Wankel engine

Rotary engine of the eccentric rotor type. Only two primary moving parts are present, the rotor and eccentric shaft. The rotor moves in one direction around a trochoidal chamber which contains intake and exhaust ports.

Ward-Leonard System

Method of regulating the speed and direction of rotation of a d.c. motor by varying its armature voltage by controlling the field current of a d.c. generator supplying the motor armature, giving a wide range and fineness of control in a low current circuit. Used on anchor capstans and mine winding gear but is high in capital cost.

Warps

Ropes used to haul ship into position when docking.

Wash

Waves caused by passage of a vessel.

Waste heat

Heat arising from the combustion of fuel in a heat engine which is dissipated without doing useful work. Heat passing out with exhaust gas and cooling water are examples.

Waste-heat boiler

With the diesel engine is usually associated a boiler in which steam is raised using the engine exhaust gases.

Water gauge glass

A glass tube connected to the outside of a vessel containing water to indicate visually the level of water within the vessel.

Water guard

Member of Customs and Excise marine anti-smuggling service.

Water hammer

Percussion in water pipe when tap is turned off or in steam pipe when live steam is admitted.

Water jet propulsion

The main engine drives a seawater pump, instead of a propeller, ejecting a jet of water astern. Jet propulsion is less susceptible to underwater damage than a propeller particularly for inshore operation and is safer if divers or swimmers are involved.

Water pocket

(1) A collecting place for water from which it can be drained, e.g. in an air compressor intercooler. (2) A collecting place for water as a result of bad design, which will lead to subsequent corrosion.

Water pressure test

(1) The testing of a tank, bulkhead, etc. by filling to the maximum working head with water. No leaks must occur, deformation may or may not be permitted depending upon the classification of the tank or bulkhead, etc. (2) The testing of a pressure vessel or pipe system by hydraulic pressure to the test pressure laid down by the regulations which may be several times the working pressure.

Waterplane area coefficient (Cw)

Ratio of the area of the waterplane to that of the circumscribing rectangle having a length and breadth equal to that of the ship.

$$Cw = \frac{\text{area of waterplane}}{L \times B}$$

Watertight bulkhead

See *Bulkhead*.

Watertight doors

Such doors below the waterline are either vertical or horizontal sliding type. The doors can be operated by power and actuated by remote control.

Watertube boiler

Steam boiler in which the heating surface consists almost entirely of an array of tubes connecting a steam drum at the top with one or more water drums, or headers at the bottom, the arrangement being such as to promote circulation of the cooler water downwards, and the hottest water and steam upwards. A rapid steaming, but often not very compact type of boiler, which lends itself to the use of many kinds of fuel.

Waterwall

The sides which form the walls of modern

water-tube boilers consist of a bank of tubes encased in refractory material. The tubes are part of the boiler water circulating system and act as downcomers.

Wattmeter

Instrument for measuring the power in watts in an electrical circuit. In one form it may work on the principle of the Siemens' dynamometer, consisting of a fixed coil of fine wire and a moving coil of thick wire. The two coils, at right angles to each other and connected in series, develop a torque between them when either direct or alternating current is passed through them, and this is measured by the torsion of a spring, the torque being directly proportional to the watts (i.e. product of amperes and volts) in the circuit.

Wave resistance

The energy wasted as a result of a ship producing waves as it moves through the water. Also known as wave-making resistance.

Wave spectra

The energy content of waves in a sea is shown in the wave spectra. The energy content of a wave is proportional to the height squared. Wave height is in turn dependent upon wind speed, the direction of the wind and the distance over which the wind has been blowing. Several formulae exist for the determination of wave spectra each taking into account the above factors in various different ways.

Waves

Oscillations of the water particles at the surface. There are three main types: ripples, translation waves and oscillating waves. The ship designer is concerned with the third type. These wave systems are created by the passage of a ship: a bow system, a stern system – both divergent – and a transverse system.

Wave winding

Distributed winding of a rotating electrical machine whose sequence of connections is such that it progresses in one direction around a machine by passing successively under each main pole of the machine.

Weardown

Wear of the main bearings of an engine

allowing the shaft to drop below its normal level.

Wear test

Specimens may be rubbed together under a given load to measure the amount of material removed from the surfaces in contact after a given time. It is an empirical test which varies in type for specific conditions.

Weather deck

Uppermost continuous deck.

Weather routeing

The use of meterological and oceanographical information to provide the most favourable route, in terms of good weather, for an ocean crossing by a ship or valuable tow.

Web

To provide extra strength and usually consisting of a plate flanged or otherwise stiffened on its edge.

Welding

Joining two metals by the application of heat and/or pressure. (1) Arc welding. Melting the metals by heat from an electric arc. (2) Gas welding. Melting by heat from a gas burning torch. (3) Resistance welding. Passing an electric current between the metals to generate heat by the resistance of the junction. (4) Friction welding. Generating heat by rotating one metal and pressing it against the other. (5) Plasma welding. Volatilization of the depositing metal in a plasma arc. (6) Electron beam welding. Melting the metals in a vacuum by focussing a beam of electrons at the joint.

Well

The open deck space between erections.

Wet liner

Removable cylinder barrel, sealed at both ends against surrounding coolant, in a cylinder block but with coolant circulating around the centre section.

Wet sump

Detachable lower half of a crankcase in which the lubricant is allowed to drain and remain until re-circulated.

Wetted surface

That part of the external hull of a ship below the deep load line.

Wheatstone bridge

(1) Instrument for measuring electrical resistances. A network of resistances as connected as in Figure 17. When the galvanometer (G) shows no deflection, the current in the four arms is balanced and R1/R2 = R3/R4. R1 and R2 are resistances of known value, R3 is a variable resistance and R4 is an unknown resistance. If R3 is then adjusted until the galvanometer shows no resistance, the value of R4 can be calculated from the above formula. (2) Technique used for monitoring flammable compounds in the atmosphere. The resistance of an active element on which controlled combustion takes place is compared with an inactive element, the pair being part of a Wheatstone bridge.

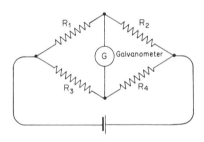

Figure 17. Wheatstone bridge circuit

Wheel house

Erection on navigating bridge deck wherein is located the steering wheel. It is also the centre for other purposes connected with the navigation of the ship.

Whip

Rope and pulley hoisting apparatus.

Whirling

Resonant transverse vibration of a rotating shaft.

Whipping

(1) Transient response caused primarily by wave impact. Fast ships having small block co-efficients and large bow flare are particularly susceptible to this type of wave induced stress. Even in the case of a hogging ship compressive bending moments can be induced in the deck. (2) Twine or similar material passed round the end of a rope to prevent it unlaying.

Windsail

Canvas funnel conveying air to lower parts of a ship. Can be used to blow out gas fumes. It is fitted with hoops in order to retain its shape.

Working

(1) With regard to materials, a movement is implied leading to opening of joints or fatigue failure of metals. (2) For machinery or manpower, some productive useful operation is taking place when the term working is used.

Worm gear

A toothed wheel worked by a revolving spiral. Single, or multi-start threads on the worm engages with suitably shaped teeth on the periphery of the worm wheel, the axes of worm and wheel generally being at right angles. Rotation of the worm pulls, or pushes the teeth of the wheel and so causes rotation. A form of gearing useful where large reduction ratios are required in a single stage; and where non-reversibility is important as when a gun mounting fires.

Woodruff key

Semi-circular key for connecting coupling to shaft. Keyway in shaft is milled by cutter of same radius as key with normal keyway in coupling or hub.

Y

Yard

(1) Cylindrical spar tapering to each end. (2) Enclosed site used for specific purpose such as shipbuilding.

Yawing

Rotation about a vertical axis is called 'yawing'. See *Swaying*.

Yield point

In a tensile test, the stress at which deformation of the test piece first occurs without any increase in the load is known as the yield point or yield stress. Not all materials exhibit a yield point in tensile testing.

Yoke

(1) Crossbar of rudder on pulling boat. (2) Coupling piece of two pipes discharging into one. (3) Main frame of an electric motor,

made of cast or rolled steel, to which field coils and motor feet are attached.

Z

Zero

Starting point in scales from which positive and negative quantity is measured.

Zinc chromate

Certain pigments in paints confer protection from corrosion on steel. Red lead and zinc chromate are good examples of such pigments. Zinc chromate is a cathodic inhibitor.

Zirconium (Zr)

Metal element (sp. gr. 6.5) used in some light alloys. The oxide, zirconia (ZrO_2) is extremely hard and refractory.